THE SKEW POLYHEDRA

BEING THE EIGHTH PART
OF SEVERAL COMPRISING
THE COMPLETE? POLYHEDRA
BY PATRICK TAYLOR

THE SKEW POLYHEDRA

First published 2013 by

Polystar Press
277 Cavendish Street
Ipswich
Suffolk
IP3 8BQ

e-mail: polystar@ntlworld.com

(01473) 434604

Made in Cornwall

ISBN 978 0 9516701 9 4

POLYSTAR PRESS IPSWICH

CONTENTS

COLOUR PLATES

ILLUSTRATIONS

Figures * are supported by Tables nos. 1-14 and 16-29

for Father Magnus Wenninger,
his ‘Polyhedron Models’ were an early inspiration

“The journey of a thousand miles starts with a single step”

Old Chinese proverb

INTRODUCTION

The geometric shapes explored in this eighth part of '*The Complete? Polyhedra*' are those formed by assembling polygons in a way which allows more than 360° worth to surround a vertex. The polyhedra themselves generally have less than 360° of polygons around each vertex, whilst the tilings have exactly that amount.

Squeezing more polygons in makes the vertices fold up into three dimensions, producing forms that have the appearance of closepacks. It must be remembered that these are actually surfaces, not solids, indeed surfaces which wind their ways in and around the solids of the closepacks. Thus whilst tilings fill the plane and polyhedra can be seen as tilings of a sphere, these skew polyhedra will be seen to be tilings of certain minimal surfaces.

Having taken the next logical step of squeezing in more polygons around a vertex to discover these forms independently for myself (as an architectural student in the early 1970's I had described them as 'space frames'), it was disappointing and yet at the same time reassuring to find that Professor Coxeter had been there many years before me with his paper outlining the basic forms.

It turns out that these three simple forms, shown overleaf in Figures 1 and 2, are but the first simple steps in a greater journey that fills the rest of this volume and promises to also fill the next.

With the experience of the seven preceding volumes in this series, the stage is therefore now set to take those basics and expand them to provide a more Complete? coverage of the many and varied possibilities of skew polyhedral forms.

REGULAR SKEW POLYHEDRA

The first three 'regular' skew polyhedra are those formed by taking the basic tilings of square and hexagon and squeezing more of each polygon around each vertex.

The vertex itself has to buckle up and down to accommodate this, and thus only admits even numbers of equal polygons around it. The vertex figures are in fact skew polygons, essentially the edges between the triangular faces of an antiprism with the top and bottom faces missing.

The resultant forms thus fold up out of the plane and fill three dimensions, but do so in a totally regular way, so that every vertex is identical and the surface produced divides three-dimensional space into two equal halves effectively inside each other.

Solids	E/V	F/V	S/V
cube	6	6 □	6
cube		(6) □	2
{4,6 |4}			
Cubic Hexaskew (540º)			
truncated octahedron	4	4 ⬡	2
truncated octahedron		(2) □	2
{6,4 |4}			
Truncated Octahedral Hexaskew (480º)			
truncated tetrahedron	6	6 ⬡	6
tetrahedron		(6) △	2
{6,6 |3}			
Truncated Tetrahedral Tetraskew (720º)			

Table 1: Regular Skew Polyhedra

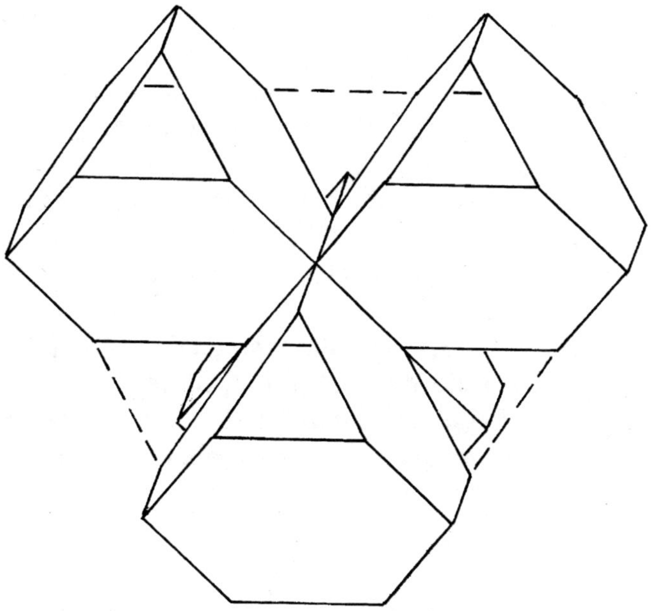

{6,6 |3}
Truncated Tetrahedral Tetraskew

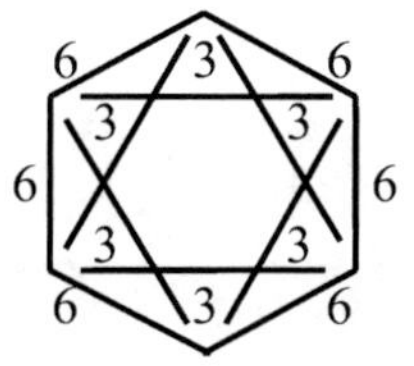

Fig.1: Self Dual Skew Polyhedron

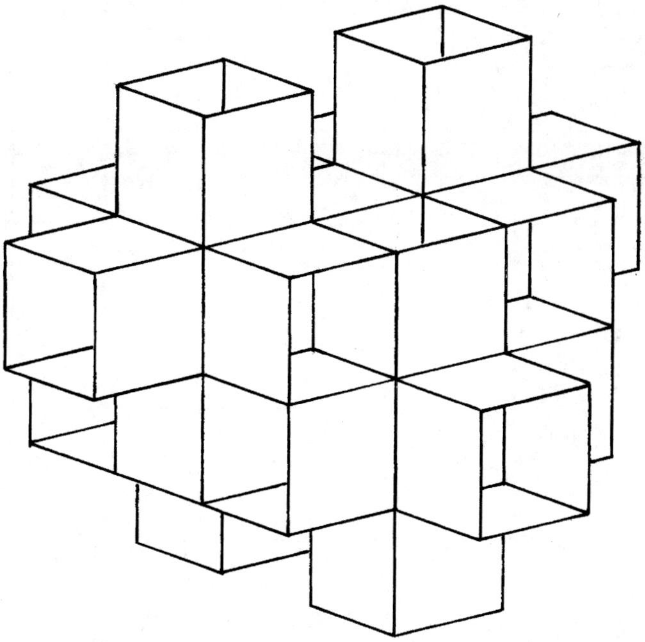

{4,6 |4}
Cubic Hexaskew

$[o^3\{4,3,4\}]$

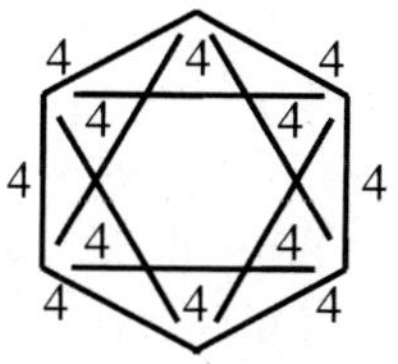

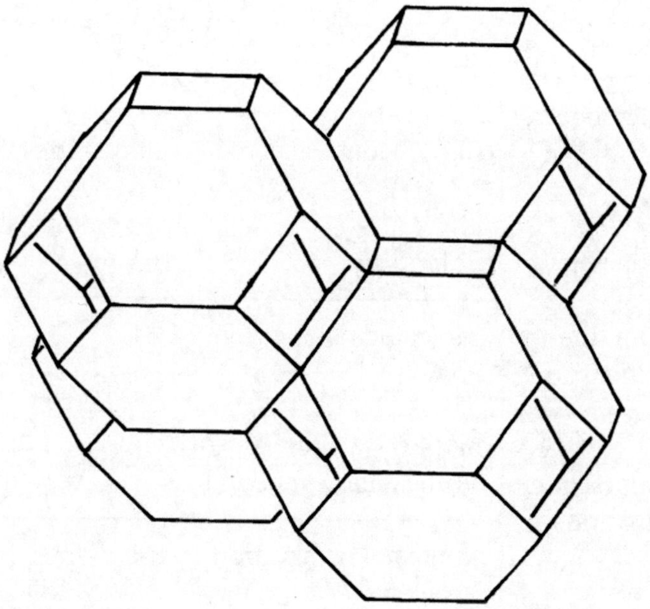

{6,4 |4}
Truncated Octahedral Hexaskew

$[t^3\{4,3,4\}]$

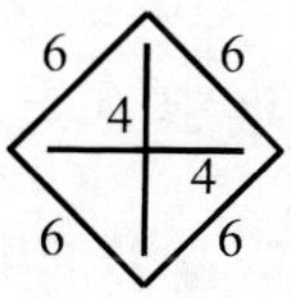

Fig.2: Dual Pair Skew Polyhedra

DUAL SKEW POLYHEDRA

Although they are just surfaces without solidity, these regular skew polyhedra can be viewed as hollow closepacks with some faces missing. Their vertex figures are similarly those of the closepacks with certain faces missing.

Six squares around a vertex would be written as {4,6} in Schläfli's notation, whilst the forms with four and six hexagons will be {6,4} and {6,6} respectively. Identifying the missing faces Coxeter adapts this notation to get {4,6 |4}, {6,4 |4} and {6,6 |3} for these first three regulars.

Schläfli's notation leads us to believe we may have here a dual pair and a self-dual. Closer examination of these regular skew polyhedra reveals that they do indeed form dual pairs like the normal polyhedra, where the centres of the faces of one are the vertices of the dual.

Starting with {6,4 |4} we can construct a series of cubes with vertices that are the centres of the hexagon faces. A single cube with all its faces missing lies at the centre of each truncated octahedron.

Other cubes with just two opposite faces missing form the tubular connections between adjoining truncated octahedra, the centres of their squares being the vertices of the hexagons.

The two skew polyhedral surfaces intersect along the diagonals of both the square and hexagon faces.

The {6,6 |3} self-dual is similar with the missing tetrahedron of one set falling centrally within the truncated tetrahedron of the other. Again the intersection of the two is along the diagonals of the hexagons.

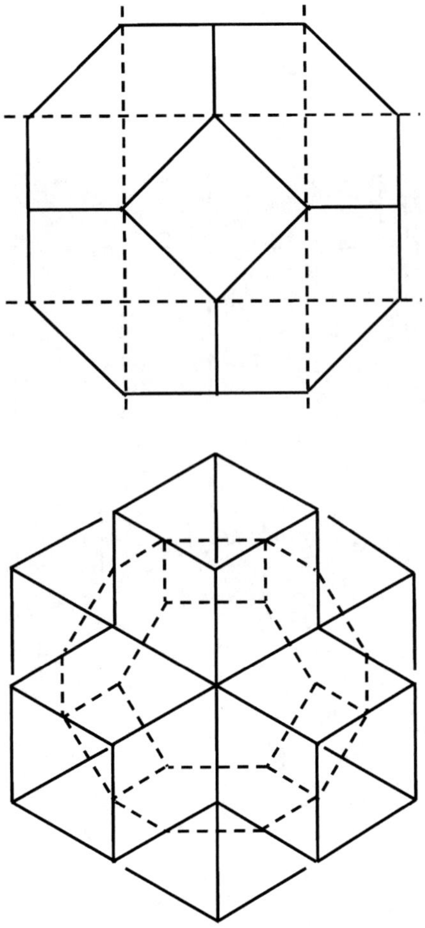

Fig.3: Dual Pair {6,4 |4} and {4,6 |4}

Solids	E/V	F/V	S/V
truncated octahedron	4	2 ⬡ 2 □ (2) ⬡	4
$nt^2\{4,6\ |4\} = nt^2\{6,4\ |4\}$ Truncated Octahedral Tetraskew (420°)			
truncated octahedron	4	2 ⬡ 2 ⬡ (2) □	4
$nt^2\{6,6\ |3\} = \{6,4\ |4\}$ Truncated Octahedral Hexaskew (480°)			

Table 2: Quasi-Regular Skew Polyhedra

TRUNCATED? SKEW POLYHEDRA

Now we have established that these regular skew polyhedra form dual pairs, the natural next question is whether there are truncation series of intermediate forms between them? In general the first truncation of a tiling or polyhedron has a triangular vertex figure, but here in the skew polyhedra such a form is not possible as the vertex figure cannot accommodate the extra angle by folding up into the third dimension.

To truncate a truncated octahedral hexaskew we would expect two dodecagons and a square at each vertex, totalling 390º. The same total angle would also be needed for the two octagons and hexagon of the 'truncated' cubic hexaskew, whilst 420º would be needed for the two dodecagons and hexagon of the 'truncated' truncated tetrahedral tetraskew.

The rectangular vertex figure of a notional second truncation for each family looks a little more promising as it can fold up into a skew polygon and thus accommodate the extra angle. For the {4,6} family this notional second 'truncation' (nt^2) would comprise two hexagons and two squares (420º). These do form a new quasi-regular skew polyhedron, as shown below. Again this comprises a continuous surface wrapped around the truncated octahedral closepack, but in a different way with hexagons missing rather than squares.

For the self dual {6,6} family the notional second 'truncation' will have two hexagons plus two hexagons (480º) at each vertex, which is simply the original truncated octahedral hexaskew {6,4 |4} we have already seen.

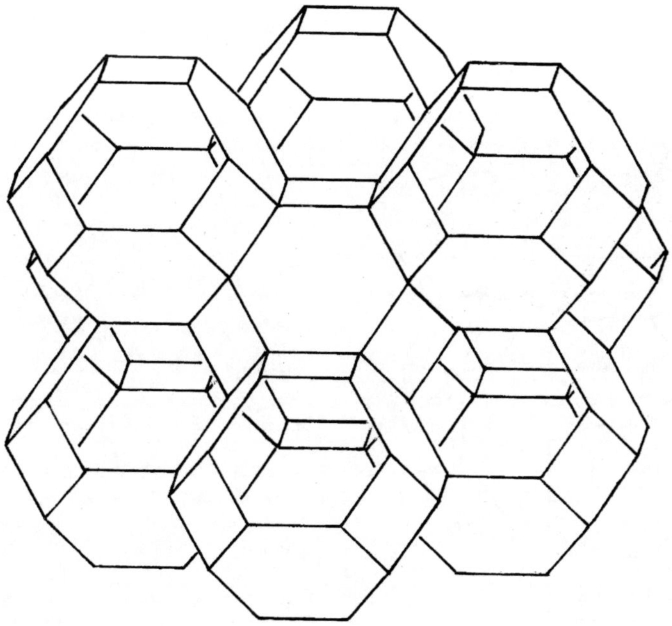

nt^2{4,6 |4} = nt^2{6,4 |4}
Truncated Octahedral Tetraskew

[o{octet}]

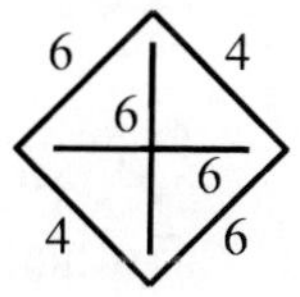

Fig.4: Quasi-Regular Skew Polyhedron

EVEN FACED SKEW POLYHEDRA

If we now take these quasi-regular skew polyhedra with their skew rectangle vertex figures as notional centres of a first truncation series, we can in fact predict a second notional truncation series.

Again the first 'truncation' is inadmissible as it would have a triangular vertex figure. However the second 'truncation' of such a second series generally has a skew trapezoid vertex figure, consisting of one of each of the parent polygons plus two squares, again an even number and therefore possible.

Constructing vertices in this way for the two symmetry families under consideration produces even faced skew polyhedra, one with one hexagon and three squares at each vertex for the dual pair and one with two hexagons and two squares for the self dual.

The latter is again the original $nt^2\{6,4|4\}$ already seen, whilst the former can be seen to be a blown up version of either of the original parent regular skew polyhedra.

Solids	E/V	F/V	S/V
trunc cuboctahedron	4	1 ⬡	2
octaprism		1 □	2
		2 □	
$nt^2t^2\{4,6|4\} = nt^2t^2\{6,4|4\}$		(2) ⯃	
Truncated Cuboctaprism Hexaskew			
truncated octahedron	4	1 ⬡	2
truncated octahedron		1 ⬡	2
		2 □	
$nt^2t^2\{6,6|3\} = nt^2\{6,4|4\}$		(2) ⬡	
Truncated Octahedral Tetraskew			

Table 3: Even Faced Skew Polyhedra

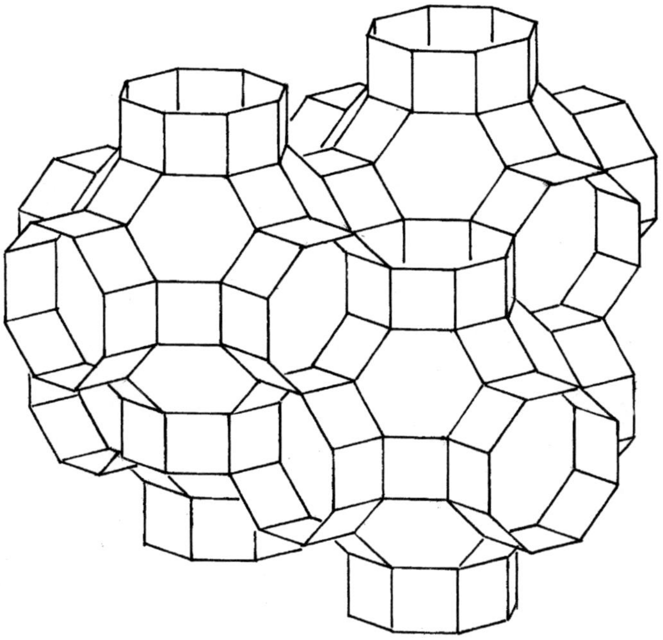

$nt^2t^2\{4,6|4\} = nt^2t^2\{6,4|4\}$
Truncated Cuboctaprism Hexaskew

[o{4,3,4}]

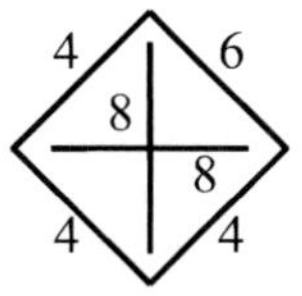

Fig.5: Even Faced Skew Polyhedron

RHOMBIC SKEW POLYHEDRA

Continuing with our notional truncation of the quasi-regular skew polyhedra we can also predict the end of the second truncation series for each family.

Fairly obviously, in line with the tilings and polyhedra, that for the self dual family will have square faces and is in fact the cubic hexaskew {4,6 |4}, the dual of the {6,4 |4} at the start of the second series.

The rhombic form for the dual pair family is a little more interesting and can be constructed by forming vertices at the centres of the hexagons and squares of the quasi-regular nt^2{4,6 |4}. As shown below,the faces of this are rhombs, in fact those of the rhombic dodecahedron, which fortunately does form its own closepack.

At this point an interesting feature of these assorted skew polyhedra should be noted.

The ends of the first series and the middle of the second series for the dual pair {4,6} family are all 6-connected with links basically arranged around a cubic grid.

The middle of the first series and its rhombic dual at the other end of the second series are both 4-connected, with links of a tetrahedral nature like the atoms of carbon in diamond.

The self dual {6,6} family is the opposite, with the ends of the first series and middle of the second series 4-connected, whilst the middle of the first series and its rhombic (square faced) dual are both 6-connected.

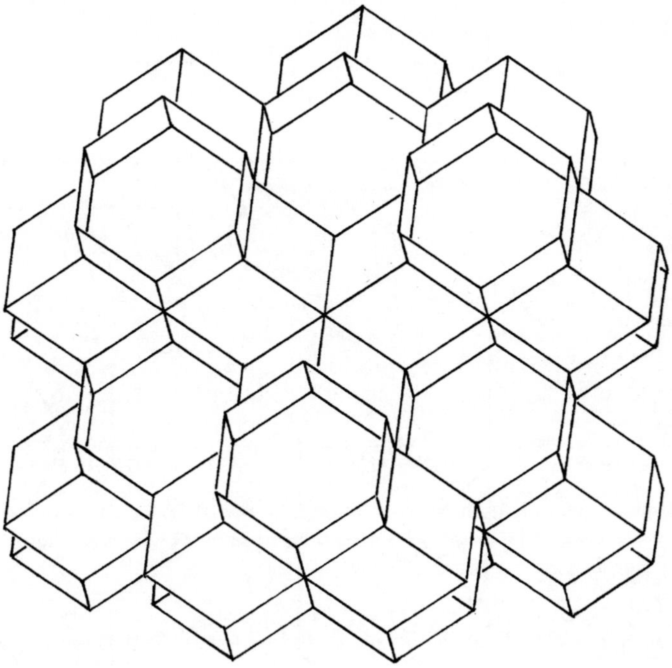

nt^4t^2{4,6 |4} = nt^4t^2{6,4 |4}
Rhombic Dodecahedral Tetraskew [t^6{octet}]

Fig.6: Rhombic Skew Polyhedron

VERTEX TRANSFORMATION

If we now take a fresh look at our two families, {4,6} and {6,6}, it would appear that notional truncation does not adequately seem to link together the various forms encountered.

However, the differing connectivities they present and the similarities between similarly connected skew polyhedra seem to provide a firmer footing on which to further investigate the relationships between the various skew polyhedra we have so far seen.

Looking more closely at the even faced skew polyhedra just described, we have already observed that that for the dual pair {4,6} family appears to be a blown up version of either of the parent regular skew polyhedra.

Taking a small section of the surface of this even faced skew polyhedron, it can be seen that to get from this to the original {6,4 |4}, we need to remove all the square faces of the {4,6 |4} and the squares immediately adjoining them. This will leave us with just four hexagons at each vertex.

This can be achieved in two stages, each involving the merging of two adjoining vertices into a single vertex by the process of transforming the shared faces. Figure 7 adjoining shows this process in terms of changes to both polygon faces and vertex figures.

The shared faces are in both these instances squares {4} and missing octagons {8}, and at each transformation these are changed to digons {2} and missing squares {4}. In general this transformation halves the value of the polygons concerned, effectively a reverse process to truncation, and whenever digons result they simply appear as edges.

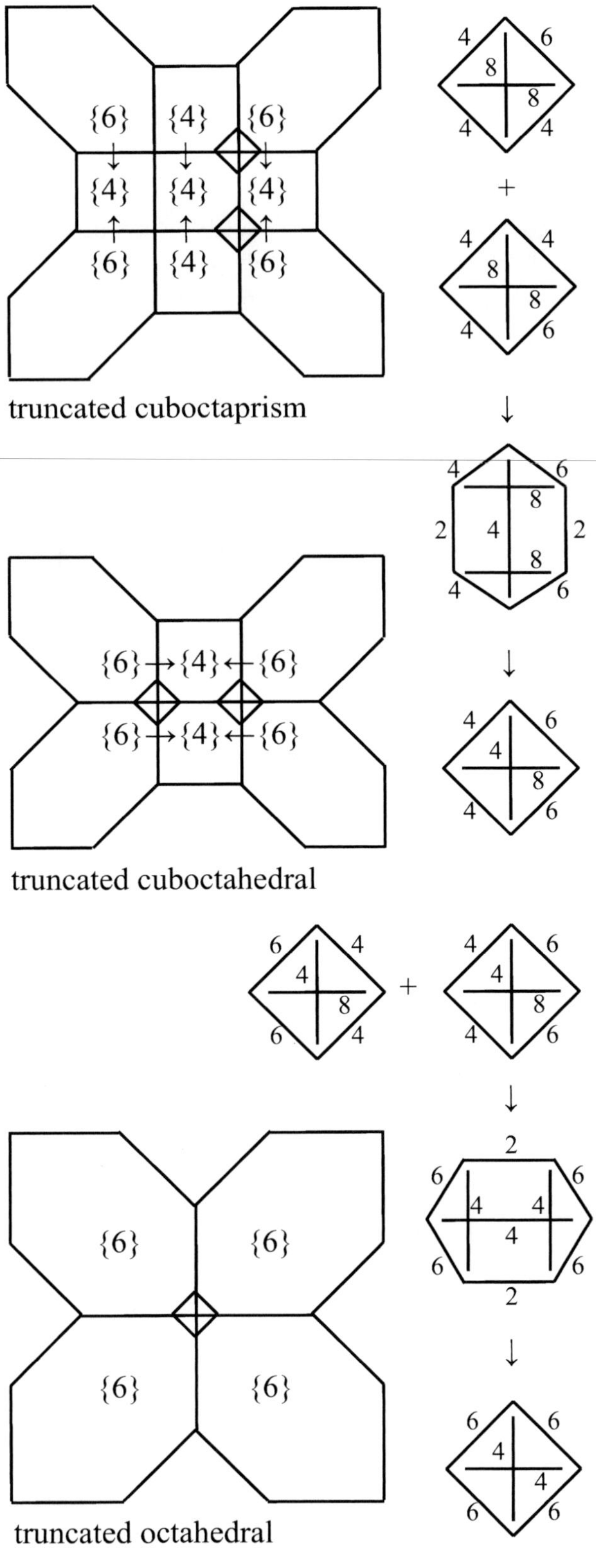

Fig.7: Transforming Skew Polyhedra

INTERMEDIATE SKEWS

If we now take another small section of the surface of this even faced skew polyhedron, it can be seen that to get from this to the original {4,6 |4}, we need to remove all the hexagon faces of the {6,4 |4} and the squares immediately adjoining them.

This will leave just six squares to each vertex. Again this can be achieved in two stages, each involving the merging of adjoining vertices into a single vertex by the process of transforming the shared faces, as shown in Figure 8 adjoining.

The shared faces in the first instance are squares {4}, hexagons {6} and missing octagons {8}, which this transformation changes to digons {2}, triangles {3} and missing squares {4}. The second transformation requires the merging of three adjacent vertices into one. The triangles disappear and the shared squares and missing octagons again become digons and missing squares.

In both the preceding transformations from the truncated cuboctaprism hexaskew towards either of the parent skews {4,6 |4} or {6,4 |4}, there has been an intermediate stage where both the targeted polygons and the adjoining squares have been reduced to polygons of half their original value, i.e. triangles or digons.

The same intermediate forms would also have resulted had our target polygons been the actual adjoining squares, those that are not part of either parent skew polyhedron. Figure 9 overleaf shows the continuation of this process with these adjoining squares as targets for removal, changing them further from digons to simple vertices. This presents another resultant skew polyhedron which can be reached from either of the intermediate ones.

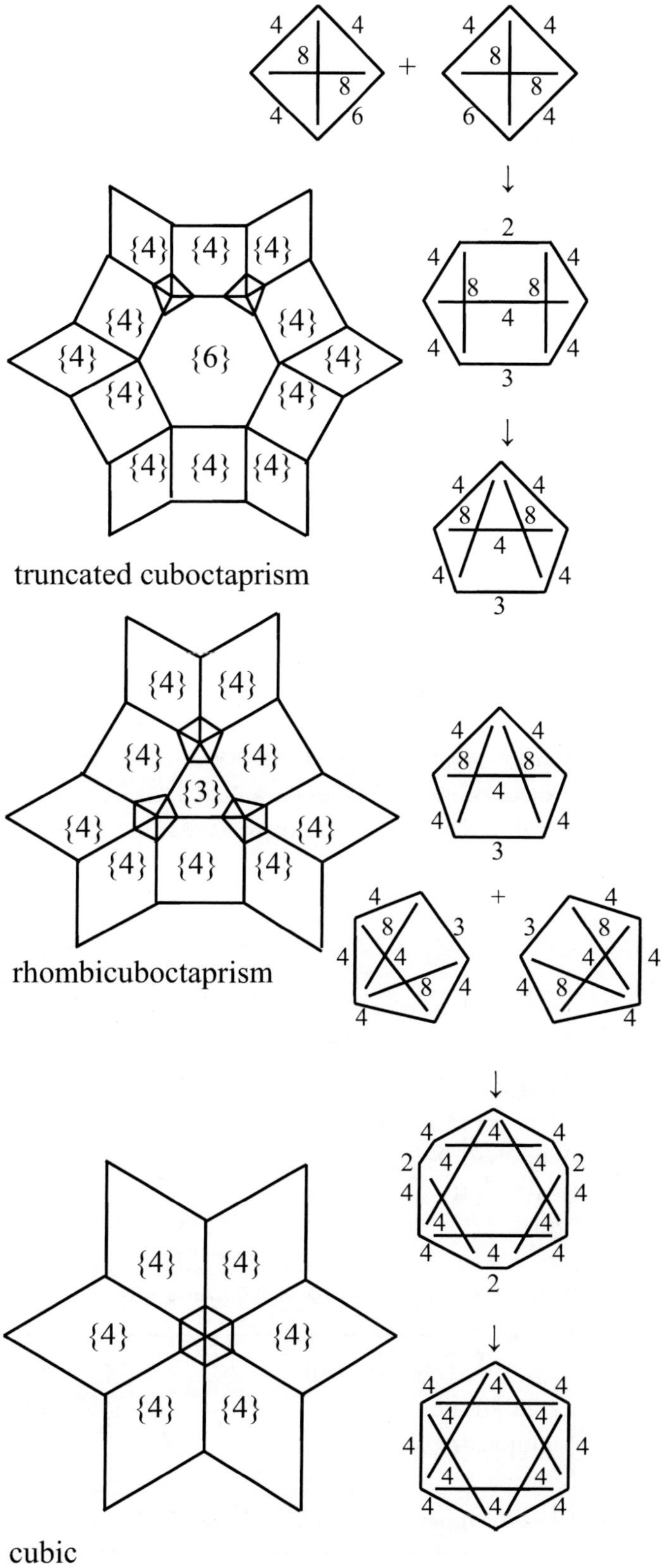

Fig.8: Transforming Skew Polyhedra

RESULTANT SKEW POLYHEDRA

This process of merging vertices to transform a skew polyhedron and find new related forms will be central to what follows.

Figure 10 facing shows the two intermediate skew polyhedra for the {4,6} family, whilst Figure 11 following shows the resultant with the 'adjoining' squares completely removed. These three new skew polyhedra in the {4,6} family do still have vertices consisting of skew polygons and present the same faces at every vertex. The surfaces produced are still complete, dividing three dimensional space into two 6-connected spaces, however these two spaces are now different, comprising different polyhedra.

The common property of all this {4,6} family is that they are 6-connected, and thus named 'hexaskews' as opposed to the 'tetraskews' of the 4-connected {6,6} family which we will now investigate.

Solids	E/V	F/V		S/V
trunc cuboctahedron	4	2	⬡	2
truncated octahedron		2	□	1
cube		(1)	[octagon]	1
		(1)	□	
Truncated Cuboctahedral Hexaskew				
rhombicuboctahedron	5	4	□	1
truncated cube		1	△	1
octaprism		(2)	[octagon]	2
cube		(1)	□	1
Rhombicuboctaprism Hexaskew				

Table 4: Intermediate Skew Polyhedra

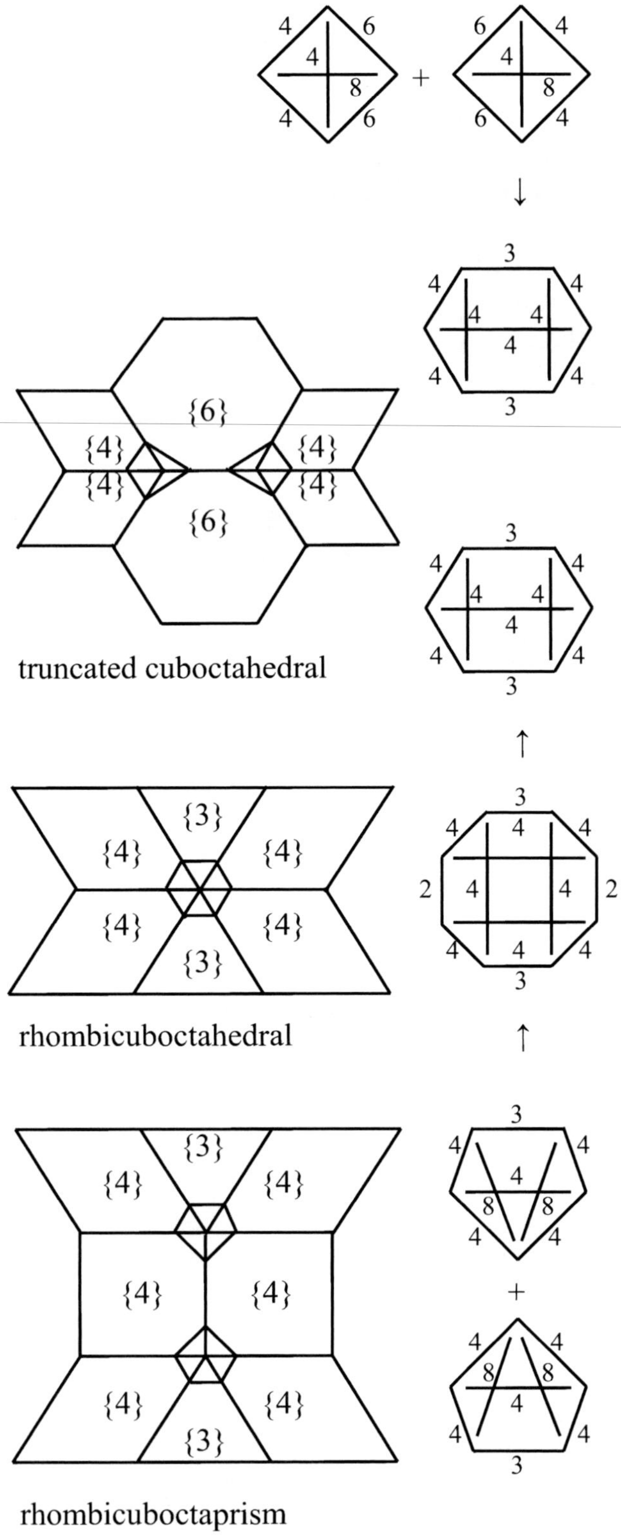

Fig.9: Transforming Skew Polyhedra

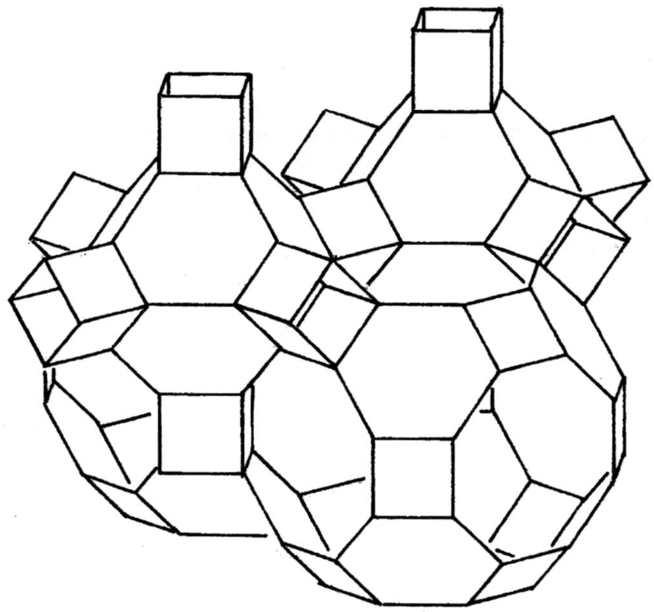

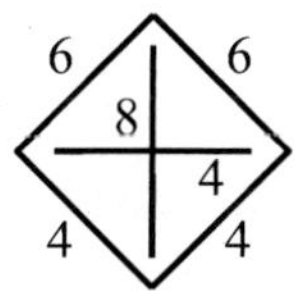

Truncated Cuboctahedral Hexaskew [tt^2{4,3,4}]

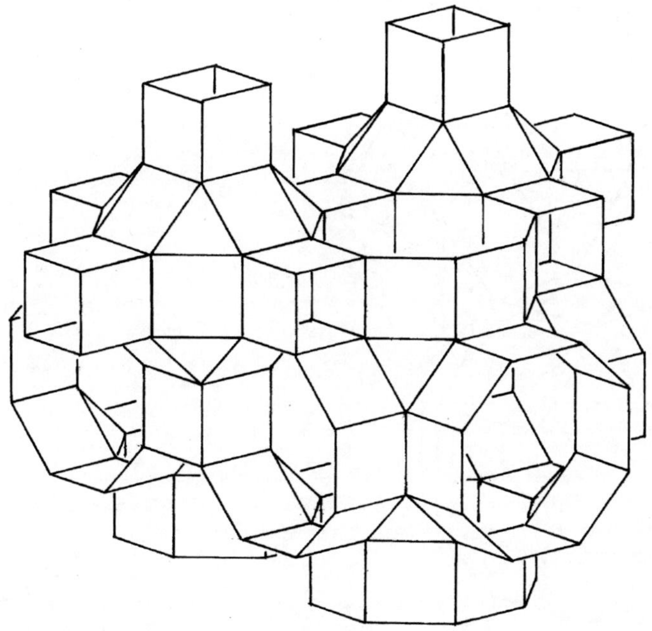

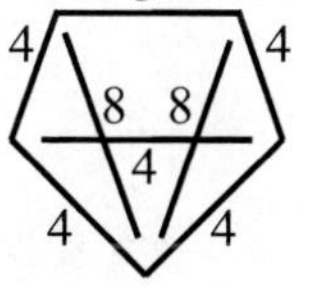

Rhombicuboctaprism Hexaskew [o^2{4,3,4}]

Fig.10: Intermediate Hexaskews

TWO MORE TETRASKEWS

The obvious question facing us now is whether the even faced skew polyhedron for the {6,6} self dual family can also be transformed this way via intermediates to produce the parent {6,6 |3}, and in turn whether the intermediates can produce another different resultant skew polyhedron.

The answer to both these questions is fortunately yes, however because of the increased symmetry within the vertex figure we are starting with, there is only one intermediate and one resultant not yet described.

These two new skew polyhedra are shown adjoining in Figure 12, and can be seen to consist of an intermediate with different polyhedra in each 4-connected space and a resultant with both spaces the same again.

Solids	E/V	F/V	S/V
cuboctahedron	5	2 ⬡	1
truncated tetrahedron		2 □	2
truncated octahedron		1 △	2
		(2) ⬡	
		(1) △	
Truncated Octetrahedral Tetraskew			
cuboctahedron	8	4 □	4
octahedron		4 △	2
		(4) △	
Cuboctetrahedral Tetraskew			
rhombicuboctahedron	6	4 □	2
cuboctahedron		2 △	1
cube		(3) □	2
Rhombicuboctahedral Hexaskew			

Table 5: More Skew Polyhedra

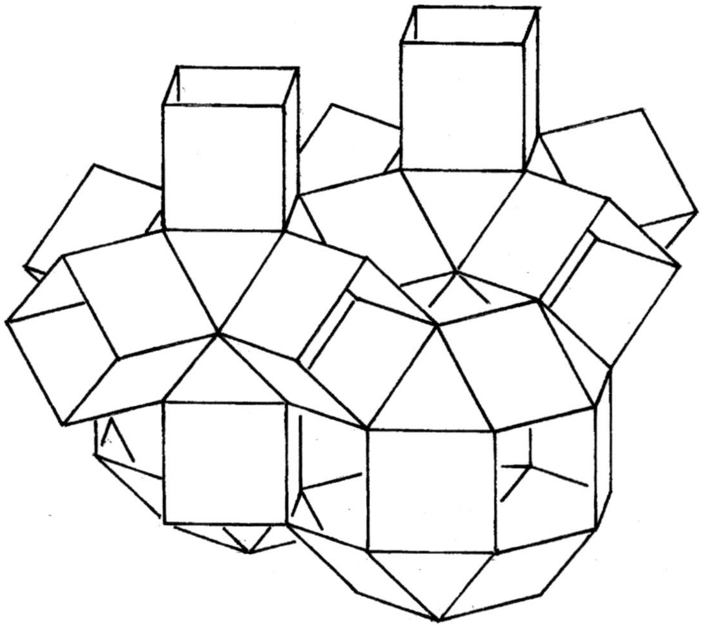

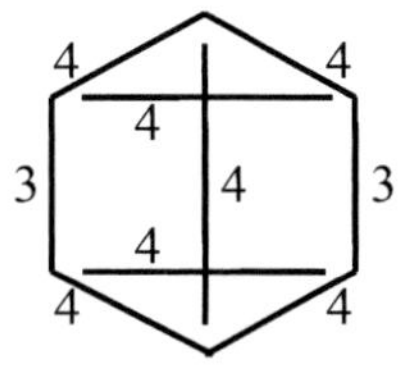

Rhombicuboctahedral Hexaskew $[t^2t^2\{4,3,4\}]$

Fig.11: Resultant Hexaskew

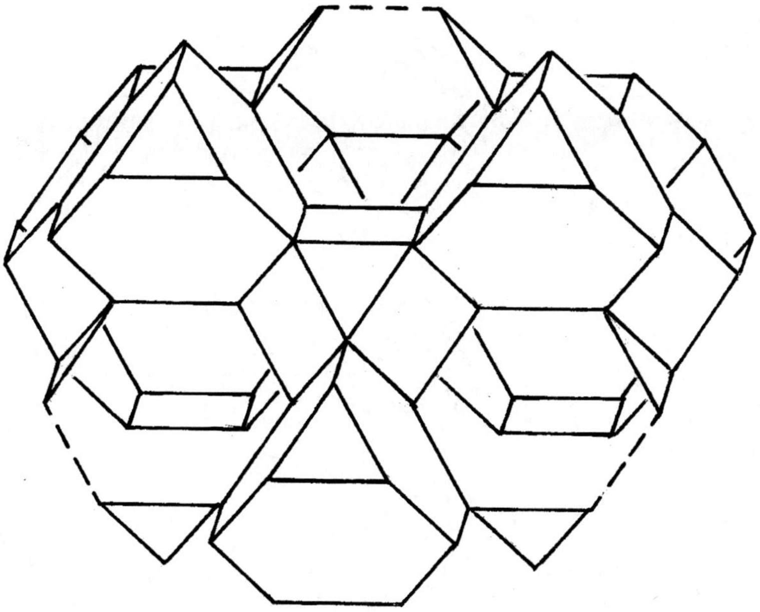

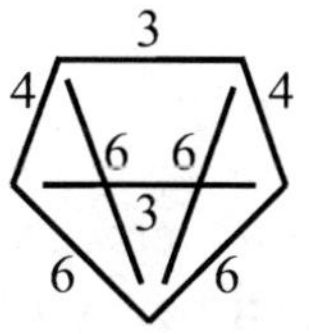

Truncated Octetrahedral Tetraskew [t{octet}]

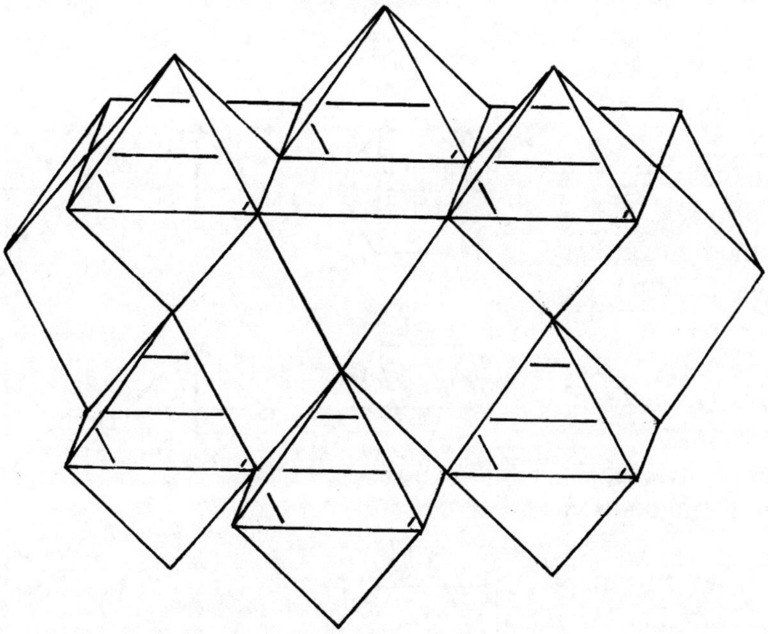

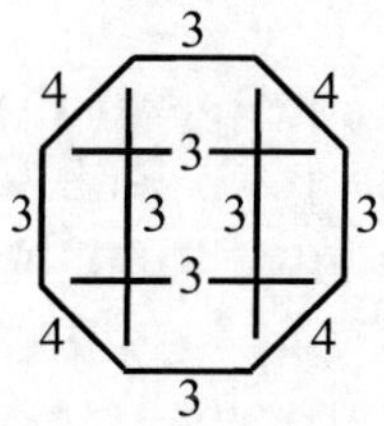

Cuboctetrahedral Tetraskew $[t^2\{octet\}]$

Fig.12: Two More Tetraskews

LINKS TO THE CLOSEPACKS

The transformation of vertices we have been examining thus far all start with the tetrahedral vertex figure of an even faced skew polyhedron. The results lead themselves naturally into an arrangement of inter-related forms within a three by three grid as shown adjoining in Figures 13 and 14 for the two families {6,6} and {4,6} respectively.

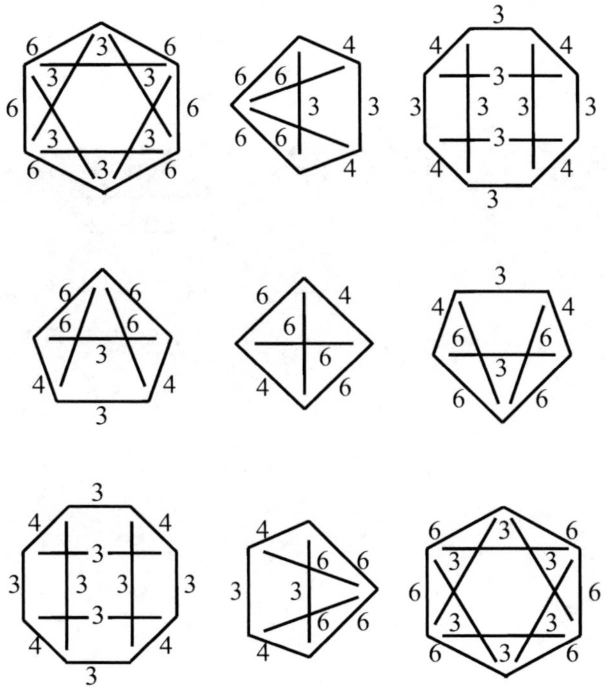

Fig.13: {6,6} Tetraskew [4-4,4-4] Grid

Each skew polyhedron produced so far relates directly to one of the closepacks, a property which has been helpful in providing them with names. It is also encouraging that the new method of transforming vertices has not thrown up any new closepacks not previously described.

In general the skew polyhedra of the {4,6} dual pair family fit around the closepacks of the {4,3,4} family and those of the {6,6} self dual family fit around closepacks of the {octet} series. In the preceding pages the relevant closepacks for each family member have been shown in square brackets [ ].

Whilst these grids are instructive, some photographs of models of the corresponding skew polyhedra, laid out on the same grid as shown opposite in Plates 1 and 2, help bring out the inter-relationships.

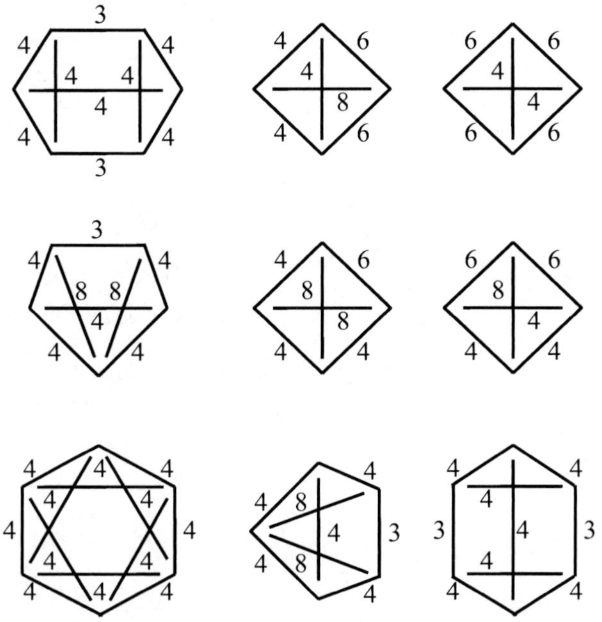

Fig.14: {4,6} Hexaskew [2-6,2-6] Grid

Whilst the {4,6} family grid is symmetrical along one diagonal, like its central member's vertex figure, that for the {6,6} family can be reflected on both diagonals.

Some intermediate and resultant forms of the {6,6} family that duplicate are thus omitted from the lower left hand corner of Plate 1 to allow space to show the rhombic dodecahedral tetraskew, dual of the central member of this grid, which shares the same 4-connected division of space.

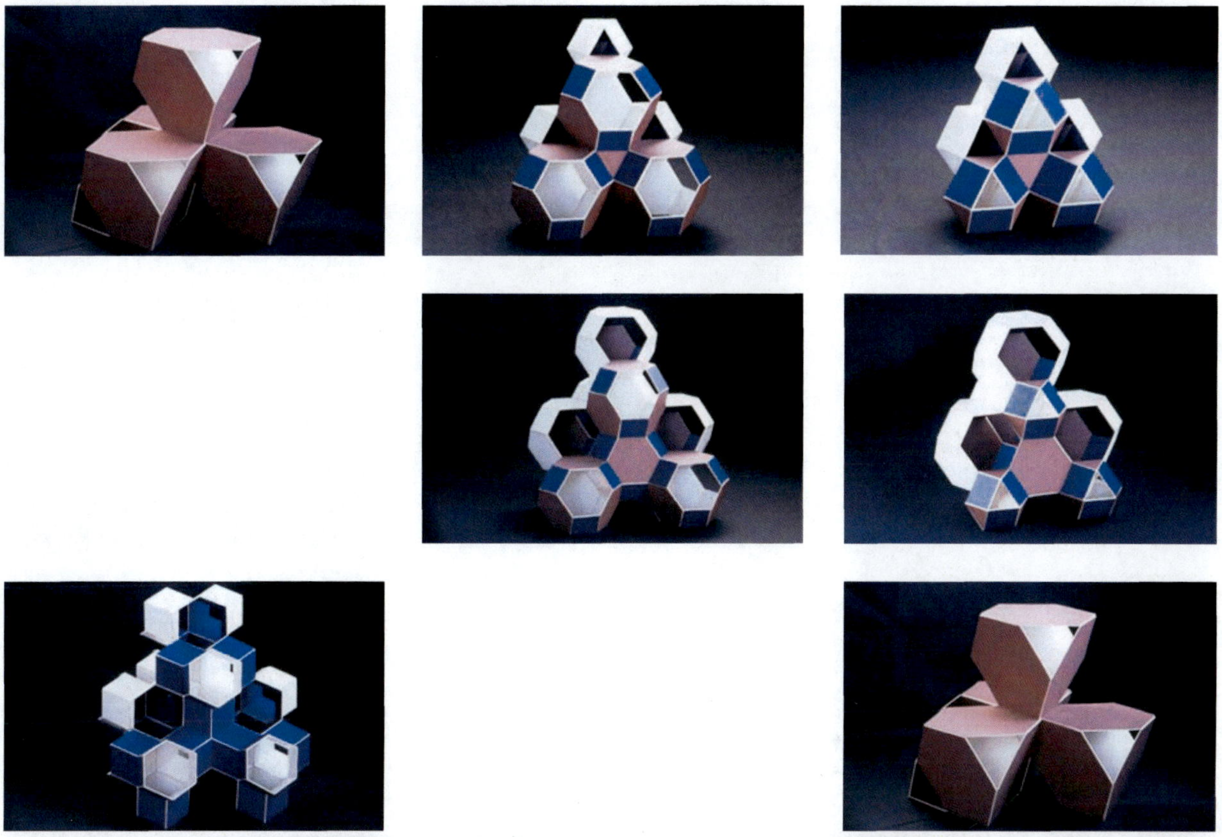

Plate 1: {6,6} Tetraskew Family [4-4,4-4]

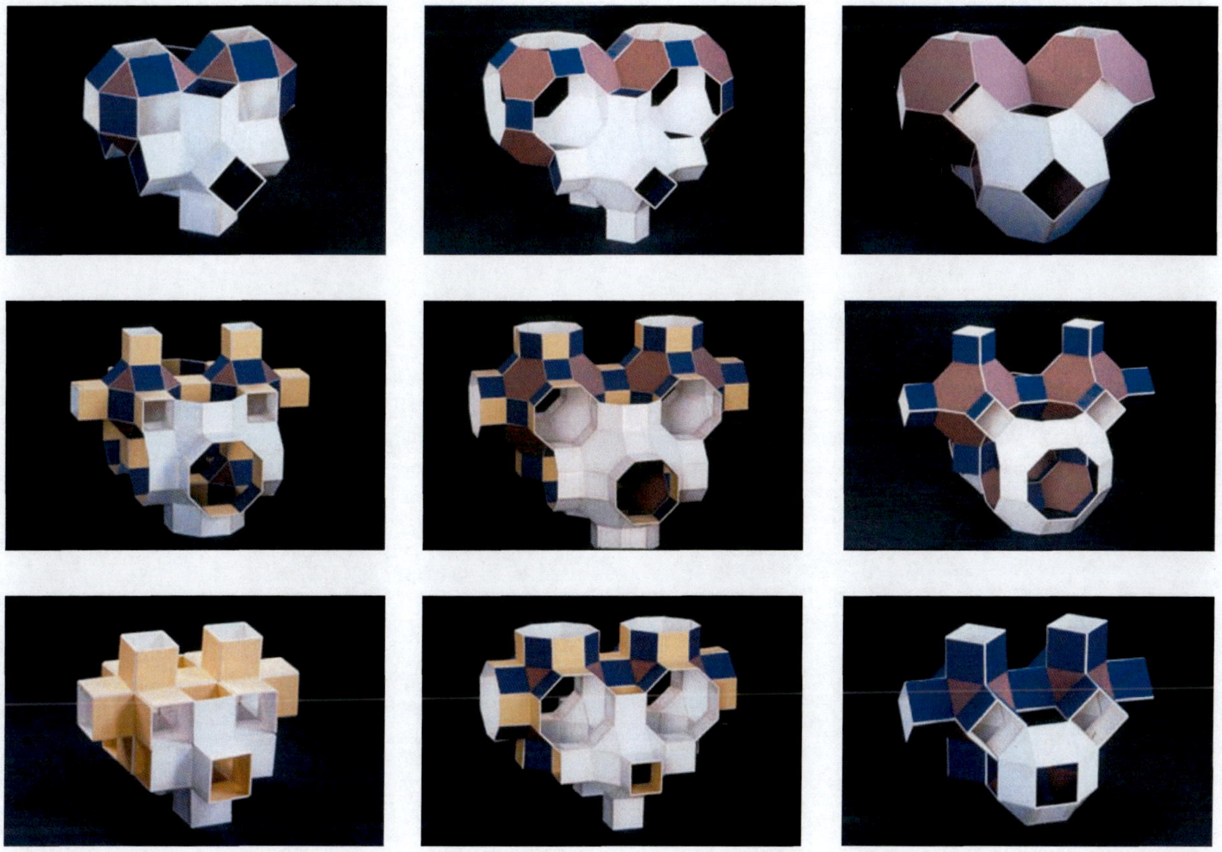

Plate 2: {4,6} Hexaskew Family [2-6,2-6]

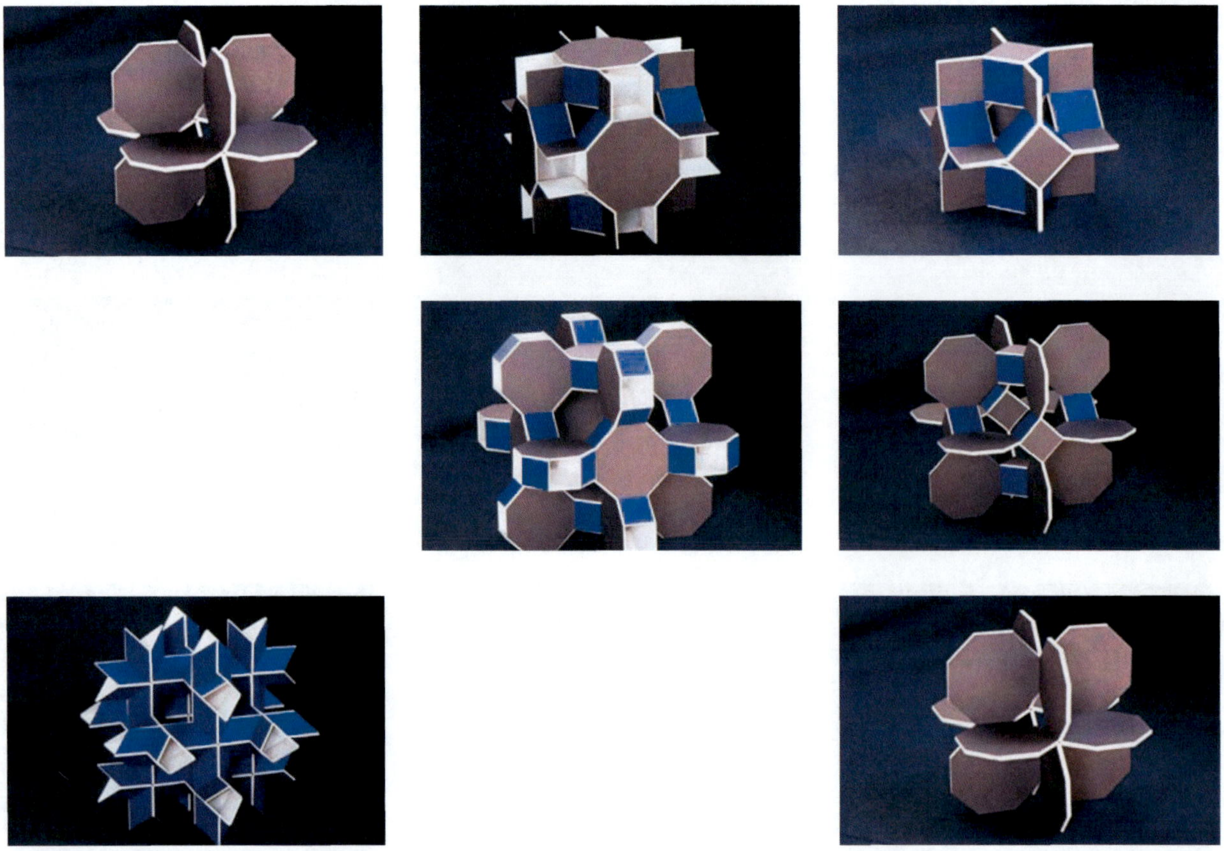

Plate 3: {8,8} Octaskew Family [4-4,8-8]

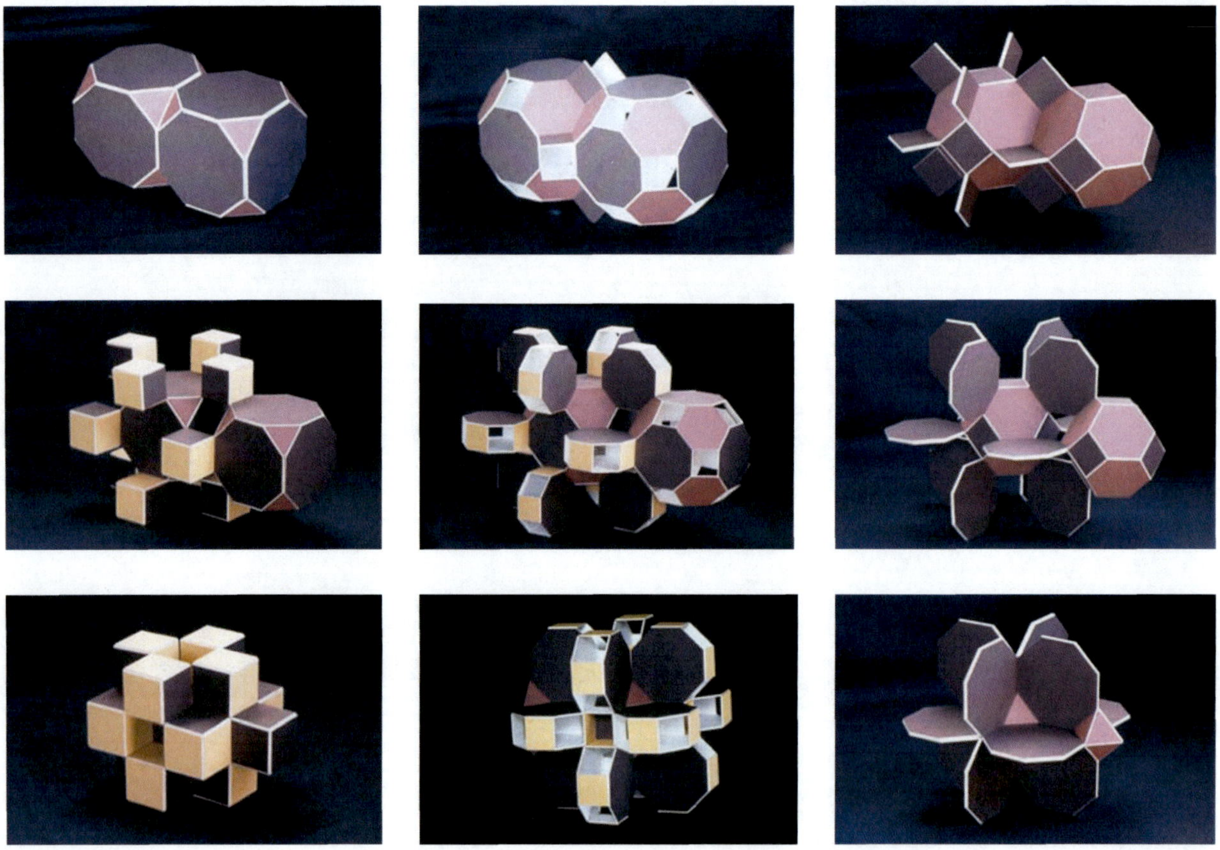

Plate 4: Dodecaskew Family [4-12,4-12]

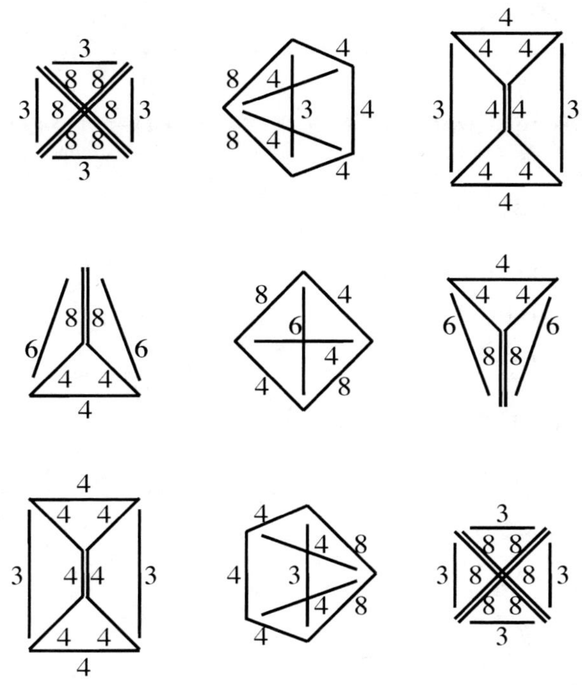

Fig.15: {8,8} Octaskew [4-4,8-8] Grid

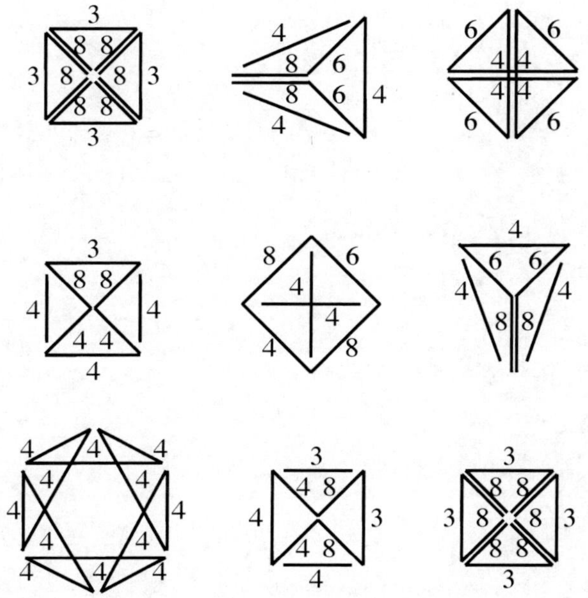

Fig.16: Dodecaskew [4-12,4-12] Grid

TWO MORE FAMILIES

Having seen that the 'intermediate' and 'resultant' skew polyhedra of the {4,6} family have connected spaces comprising different polyhedra, it is perhaps worth looking again at our starting point, the central truncated cuboctaprism hexaskew.

We can imagine a different division into two separate spaces by joining all the octaprisms together by their shared square faces, leaving the truncated cuboctahedra joined by their shared hexagon faces.

In this skew polyhedron the octaprisms are 4-connected and the truncated cuboctahedra are 8-connected. What we have actually done here is form a different skew polygon vertex figure by missing out different polygons, squares and hexagons instead of octagons, from the original tetrahedral vertex figure of the closepack.

Treating this as a starting point we can carry out vertex transformations as before to yield another family of skew polyhedra, all with both 4-connected and 8-connected spaces, which we will call the 'octaskews'.

Similarly, we can also leave out the opposing square faces from this tetrahedral vertex figure to provide a third starting point for another series of transformations. Here the central form again yields two equal spaces, each comprising both 12-connected truncated cuboctahedra and 4-connected octaprisms, and the rest of this family of 'dodecaskews' maintain these connections.

The grids for these two new families are shown adjoining in Figures 15 and 16 and their inter-relationships are well illustrated by further photographs of models of the corresponding skew polyhedra in Plates 3 and 4 opposite.

{8,8} FAMILY OCTASKEWS

Looking at the first of these families, one can in fact be regarded as a 'regular' as it has only octagons in its construction, and eight of them occur at each vertex. Fitting around the truncated cubic closepack it can be designated as {8,8 |3} in Coxeter's notation. Here the octagons are 4-connected and the truncated cubes and missing octahedra are both 8-connected.

Figure 18 shows this 'parent' regular skew polyhedron with below it the central member of the grid based on the truncated cuboctaprism closepack. Somewhat surprisingly, like the central member of the {6,6} family grid, this skew polyhedron has a dual, as shown below in Figure 17 and as the extra photograph at lower left in Plate 3. The rhombs here have angles of 74.86° and 105.14° and diameters in the ratio of the dual edge lengths of the polygons {4} and {8} as used in vertex figure construction (1:1.30656).

Solids	E/V	F/V	S/V
truncated cube	8	8 ⯃	4
octahedron		(4) △	1
{8,8 |3}		(4) ‖	
Truncated Cubic Octaskew			
trunc cuboctahedron	4	2 ⯃	2
octaprism		2 □	2
		(1) ⬡	
		(1) □	
Truncated Cuboctaprism Octaskew			

Table 6: Some Octaskew Polyhedra

Fig.17: Rhombic {4,8} Octaskew

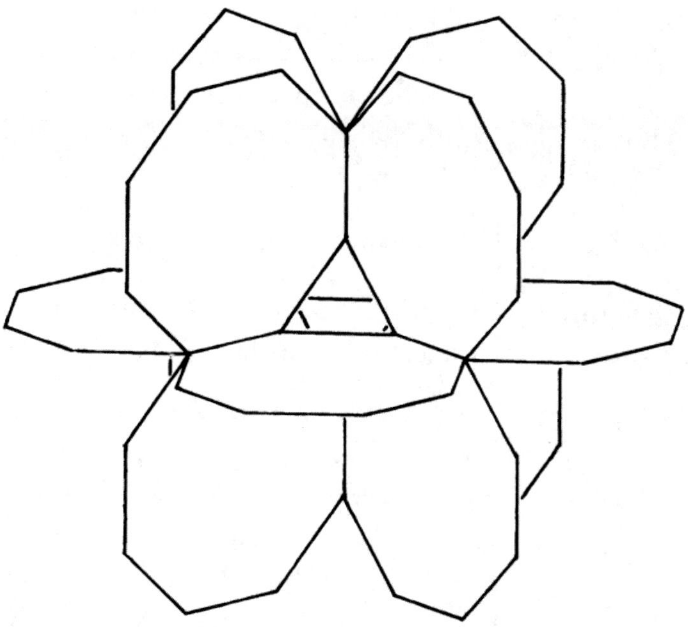

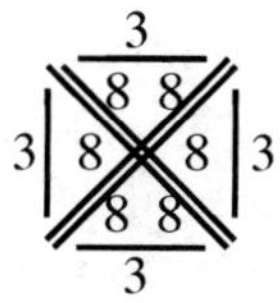

{8,8 |3}
Truncated Cubic Octaskew [t{4,3,4}]

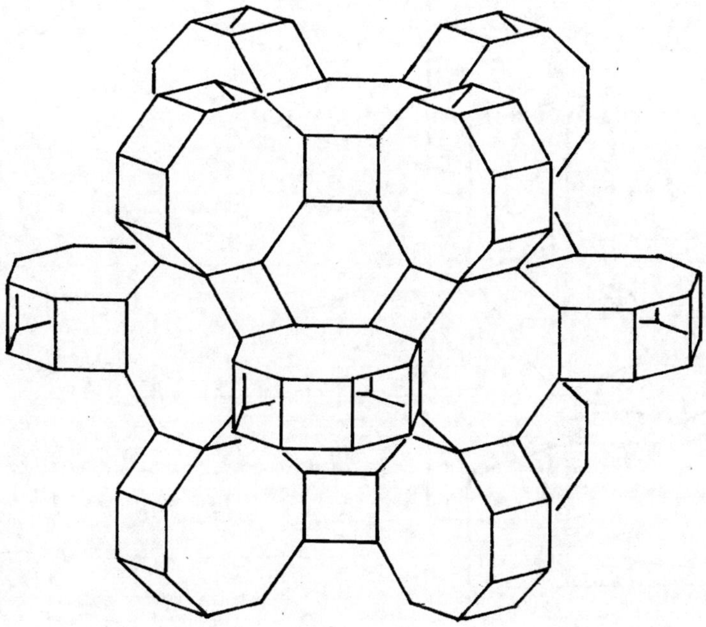

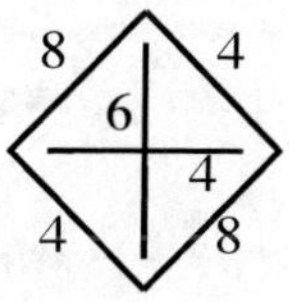

Truncated Cuboctaprism Octaskew [o{4,3,4}]

Fig.18: Octaskew Polyhedra

INTERMEDIATES & RESULTANT

The first step towards producing the parent {8,8} from the central even faced skew polyhedron can be carried out in two different ways. Reducing an octagon, square and missing hexagon to square, digon and missing triangle gives one result, whilst reducing an octagon, square and missing square to square, digon and missing digon gives another. These two intermediate forms are shown opposite in Figure 20.

Completing the process by removing the digons from either of these leads to another resultant skew polyhedron as shown below in Figure 19.

All three still relate to particular members of the {4,3,4} family of closepacks and are described adjoining in Table 7. Again because of the symmetry of our starting point, there are no further family members.

Solids	E/V	F/V		S/V
rhombicuboctahedron	5	2	⯃	1
truncated cube		3	□	1
octaprism		(2)	□	2
cube		(1)	△	1
Rhombicuboctaprism Octaskew				
trunc cuboctahedron	5	2	⯃	2
truncated octahedron		3	□	1
cube		(2)	⬡	1
		(1)	∥	
Truncated Cuboctahedral Octaskew				
rhombicuboctahedron	8	8	□	2
cuboctahedron		(2)	△	1
cube		(2)	∥	2
Rhombicuboctahedral Octaskew				

Table 7: Further Octaskews

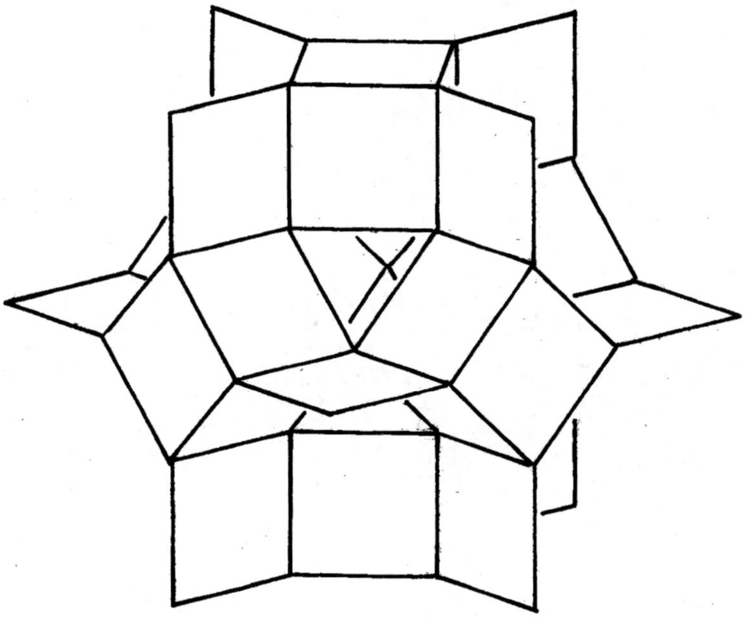

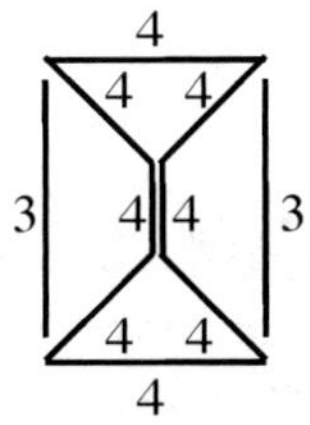

Rhombicuboctahedral Octaskew $[t^2t^2\{4,3,4\}]$

Fig.19: Resultant Octaskew

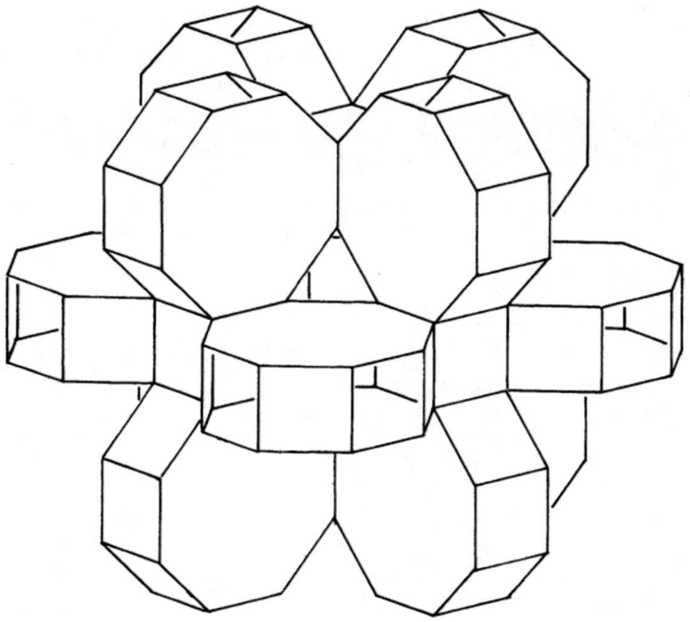

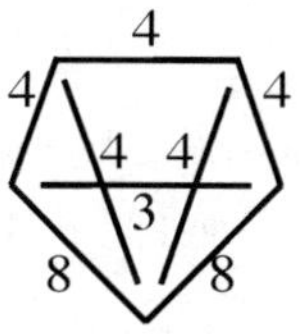

Rhombicuboctaprism Octaskew $[o^2\{4,3,4\}]$

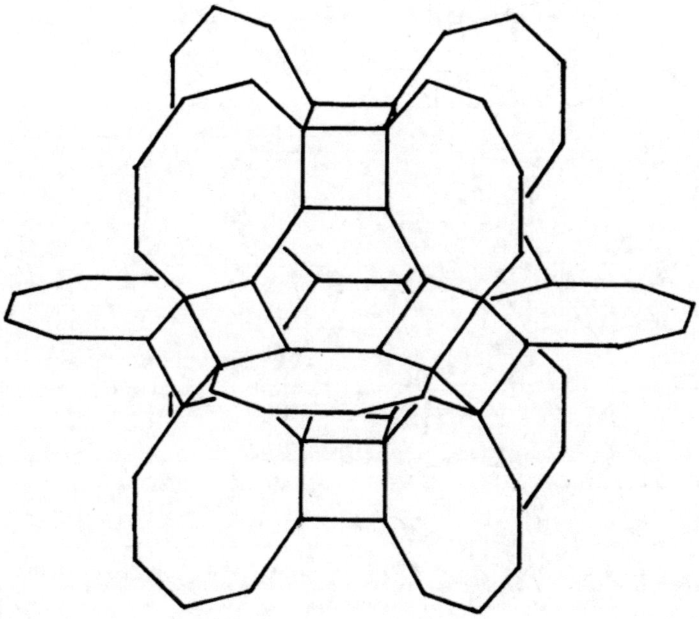

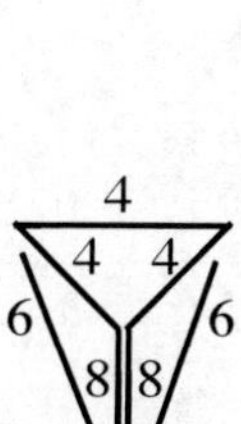

Truncated Cuboctahedral Octaskew $[tt^2\{4,3,4\}]$

Fig.20: Intermediate Octaskews

DODECASKEW POLYHEDRA

If we now turn to the second of our new families, the central member, the now third interpretation of the truncated cuboctaprism closepack, is shown below in Figure 21. It comprises two equal interlocked assemblies of truncated cuboctahedra and octaprisms.

Transforming this central member twice we can obtain two further resultant skew polyhedra that mirror the {4,6} and {6,4} parent skew polyhedra, as shown opposite in Figure 22.

These differ from the parents in that there seem to be complete polyhedra present, cubes or truncated octahedra, however the spaces within are actually 12-connected to those adjoining by missing digons providing a slot through to the next solid.

Solids	E/V	F/V	S/V
cube	12	12 □	8
		(6) ‖	
Cubic Dodecaskew			
truncated octahedron	8	4 ⬡	4
		4 □	
		(4) ‖	
Truncated Octahedral Dodecaskew			
trunc cuboctahedron	4	2 ⯃	2
octaprism		1 ⬡	2
		1 □	
		(2) □	
Truncated Cuboctaprism Dodecaskew			

Table 8: Some Dodecaskew Polyhedra

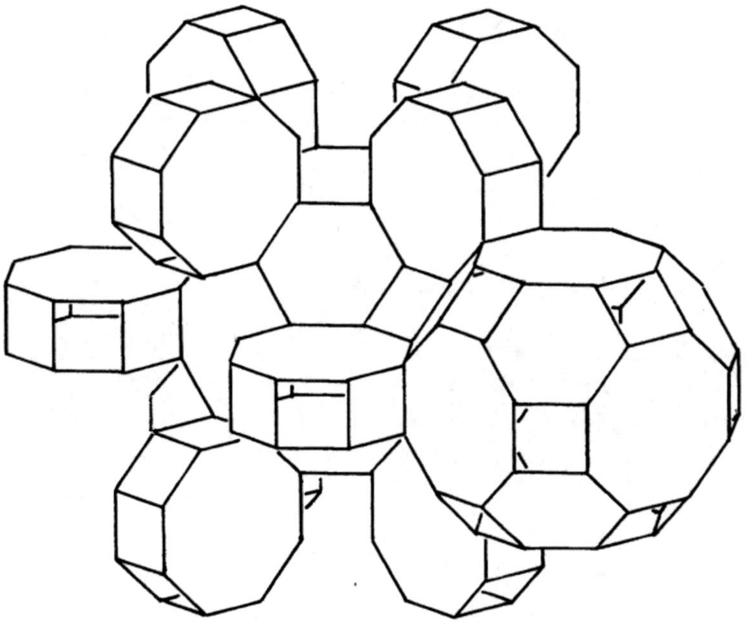

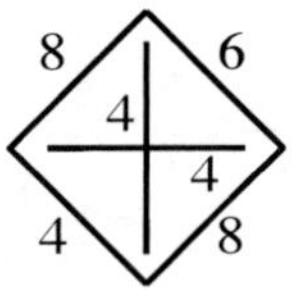

Truncated Cuboctaprism Dodecaskew [o{4,3,4}]

Fig.21: Central Dodecaskew

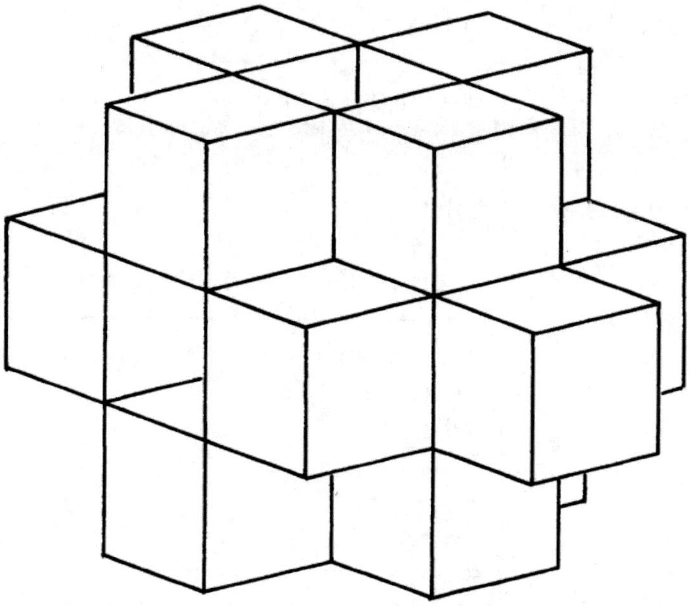

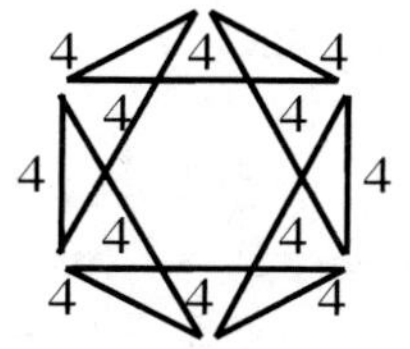

Cubic Dodecaskew $[o^3\{4,3,4\}]$

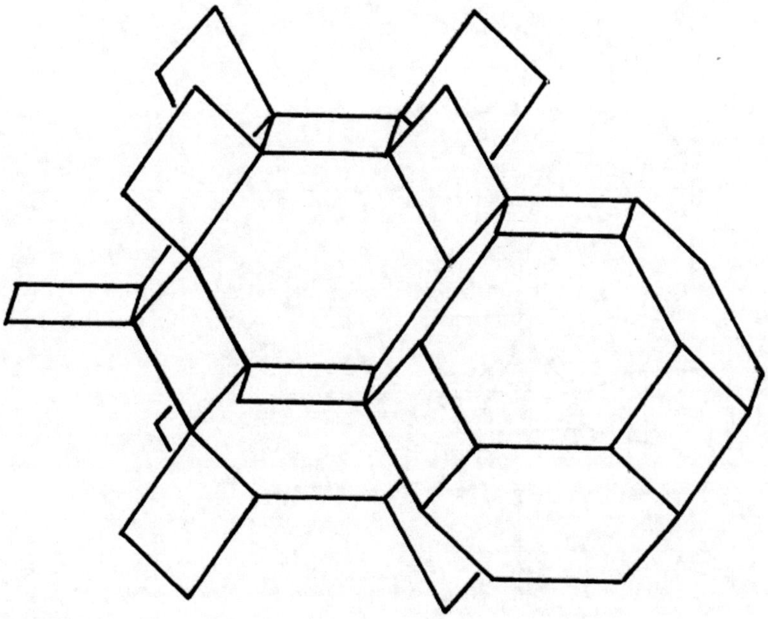

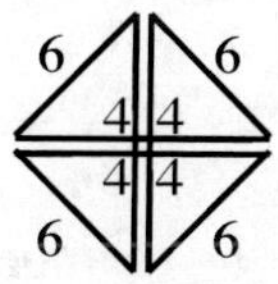

Truncated Octahedral Dodecaskew $[t^3\{4,3,4\}]$

Fig.22: Resultant Dodecaskews

INTERMEDIATES & RESULTANT

The other skew polyhedron in this family that results from two transformations also mirrors something seen before, this time the truncated cubic octaskew {8,8}. This resultant is shown below in Figure 23 and again we have no apparent holes between adjacent cells, but gaps are there in the form of missing digons. Both the truncated cubes and the octahedra are 12-connected, whilst the original edges of the truncated cubes and the octagons are 4-connected.

Opposite in Figure 24 are two intermediate forms from this family, there being no further members since the grid has a diagonal symmetry. These provide us with third versions of both the truncated cuboctahedral and the rhombicuboctaprism closepacks, this time with both 4- and 12-connected polyhedral cells.

Solids	E/V	F/V	S/V
rhombicuboctahedron	6	2 ⯃	1
truncated cube		3 □	1
octaprism		1 △	2
cube		(2) □	1
		(1) ‖	
Rhombicuboctaprism Dodecaskew			
trunc cuboctahedron	5	2 ⯃	2
truncated octahedron		2 ⬡	1
cube		1 □	1
		(2) □	
		(1) ‖	
Truncated Cuboctahedral Dodecaskew			
truncated cube	12	8 ⯃	4
octahedron		4 △	1
		(4) ‖	
Truncated Cubic Dodecaskew			

Table 9: Further Dodecaskews

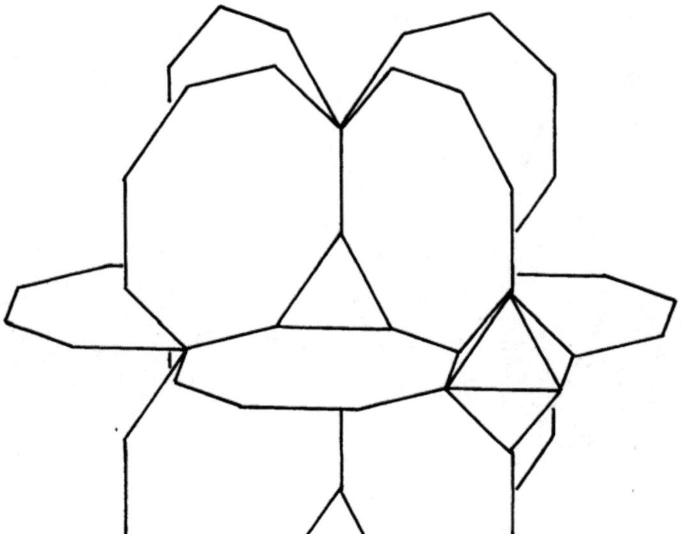

Truncated Cubic Dodecaskew

[t{4,3,4}]

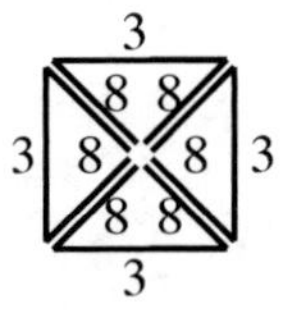

Fig.23: Resultant Dodecaskew

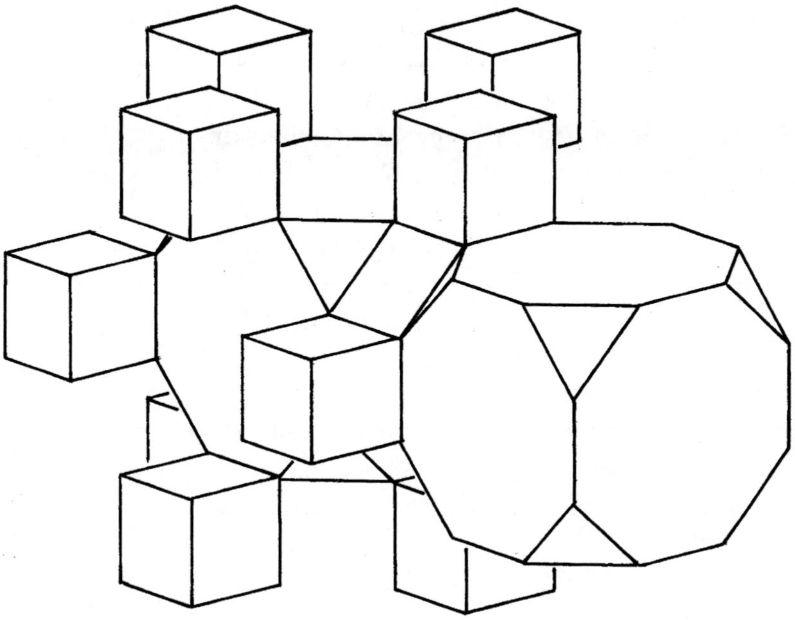

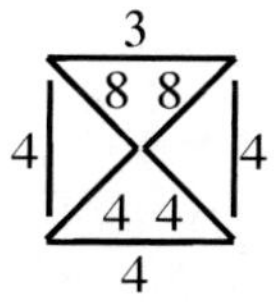

Rhombicuboctaprism Dodecaskew $[o^2\{4,3,4\}]$

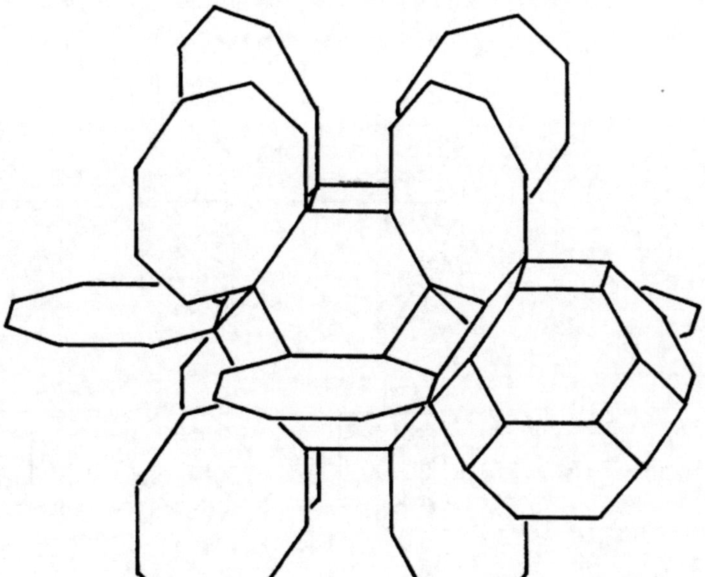

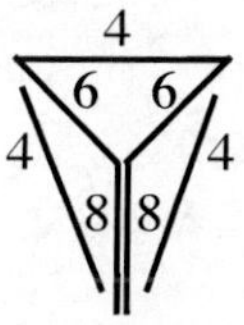

Truncated Cuboctahedral Dodecaskew $[tt^2\{4,3,4\}]$

Fig.24: Intermediate Dodecaskews

MORE HEXASKEW POLYHEDRA

The central member of the {6,6} family can be seen to be based on the first odd of the octet series: o{octet}. This can also have its tetrahedral vertex figure re-aligned in two further ways. Again because of the inherent symmetry in this vertex figure, one of these is exactly the same as the one already explored that gave us the {6,6} family of 4-connected tetraskews.

The other re-alignment however is different, producing as its central member the vertex of the parent {6,4 |4}, which generates some further 6-connected hexaskew polyhedra, two of which we have not as yet seen.

The very symmetrical grid for this family is shown adjoining in Figure 25 and the new intermediate and resultant skew polyhedra are shown opposite in Figure 26, and at top right and left respectively of Plate 5.

Solids	E/V	F/V	S/V
cuboctahedron	6	4 ⬡	1
truncated octahedron		2 △	2
truncated tetrahedron		(2) □	2
		(1) ‖	
Truncated Octetrahedral Hexaskew			
truncated tetrahedron	12	6 ⬡	6
tetrahedron		6 △	2
		(6) ‖	
Truncated Tetrahedral Hexaskew			

Table 10: More Hexaskews

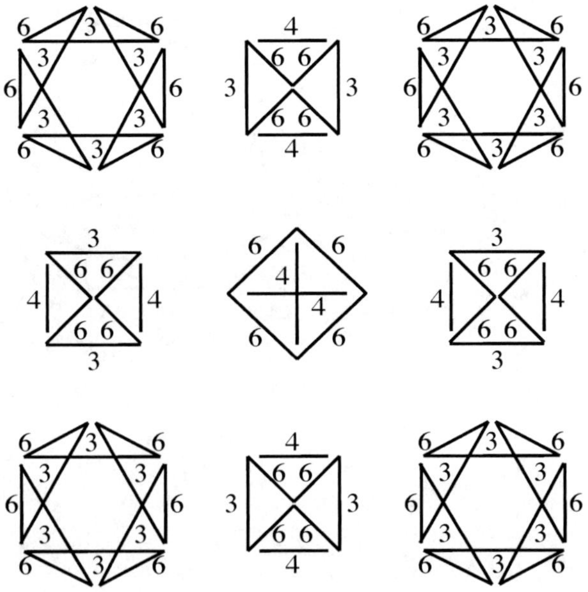

Fig.25: Hexaskew [6-6,6-6] Grid

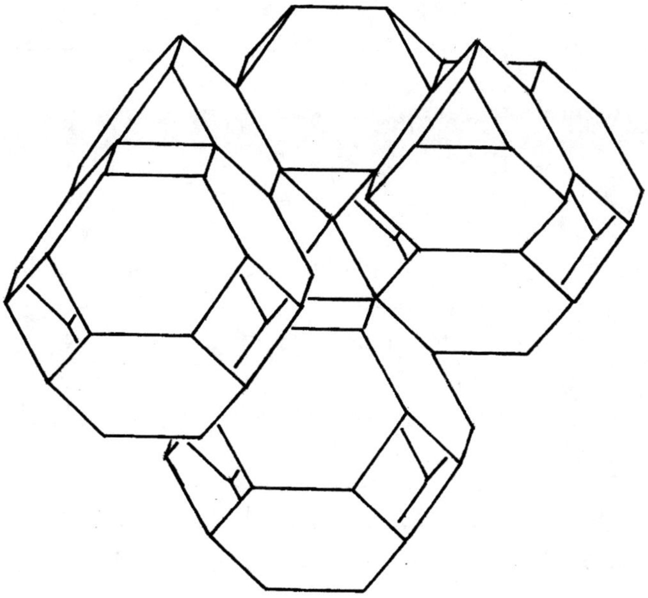

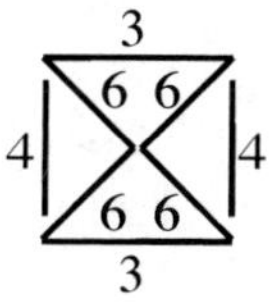

Truncated Octetrahedral Hexaskew [t{octet}]

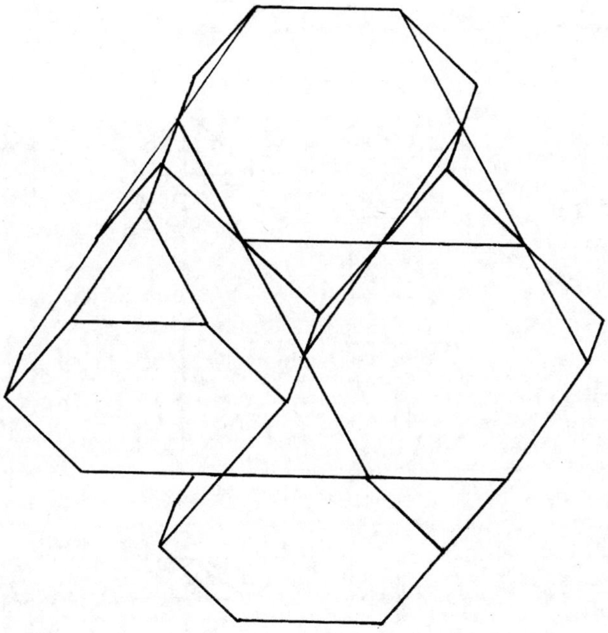

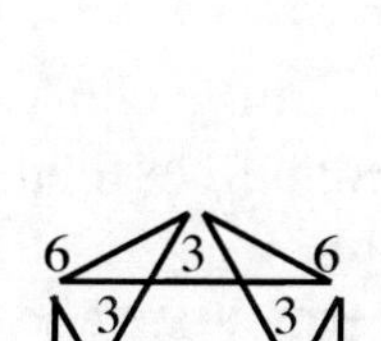

Truncated Tetrahedral Hexaskew

Fig.26: More Hexaskew Polyhedra

EVEN MORE HEXASKEWS

One further tetrahedral vertex figure was produced as an intermediate in our original {4,6} family grid, that of the truncated cuboctahedral closepack, $tt^2\{4,3,4\}$. It will be worthwhile considering this as the central member of a grid in its own right.

In one orientation it comprises two hexagons and two squares with a missing octagon and square. This produces a further family of 6-connected hexaskew polyhedra, some of which we have already seen as part of the {4,6} and {6,6} families, but two of which we have not seen before.

The grid for this family is shown adjoining in Figure 27 and the new intermediate and resultant skew polyhedra are shown opposite in Figure 28.

These are of particular interest as they are based on two underlying closepacks, which we have not seen before amongst the skews, the t^3\{octet\} and t^4\{octet\}. The latter in this incarnation comprises mainly tetrahedra, the cubes being absent.

If we now take this truncated cuboctahedral closepack vertex figure, we find that the other two alignments at the centre of a grid duplicate each other, both producing a vertex figure comprising an octagon, a hexagon and two squares around a missing hexagon and square (see Fig.36, p.41).

Whilst the octaskews and dodecaskews were all based around closepacks of the cubic {4,3,4} series, we will see that the intermediate and resultant skew polyhedra this grid produces relate to both the {4,3,4} and {octet} series. Each one still comprises two spaces, but they are very different, one 12-connected, the other with both 4- and 8-connected polyhedra, hence icosaskews.

Solids	E/V	F/V	S/V
trunc cuboctahedron	5	2 ⬡	2
truncated cube		2 □	1
truncated tetrahedron		2 △	1
		(2) ⯃	
		(1) ‖	
Truncated Cuboctetrahedral Hexaskew			
rhombicuboctahedron	9	6 □	3
cube		3 △	1
tetrahedron		(3) □	1
		(3) ‖	
Rhombicuboctetrahedral Hexaskew			

Table 11: Even More Hexaskews

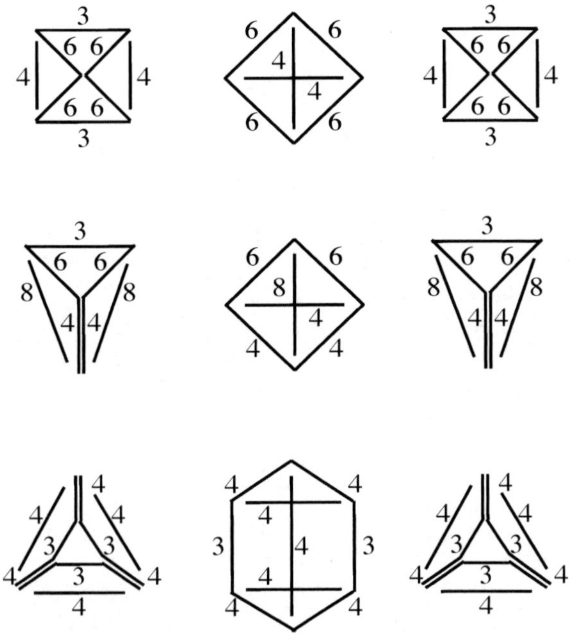

Fig.27: Hexaskew [6-6,2-6] Grid

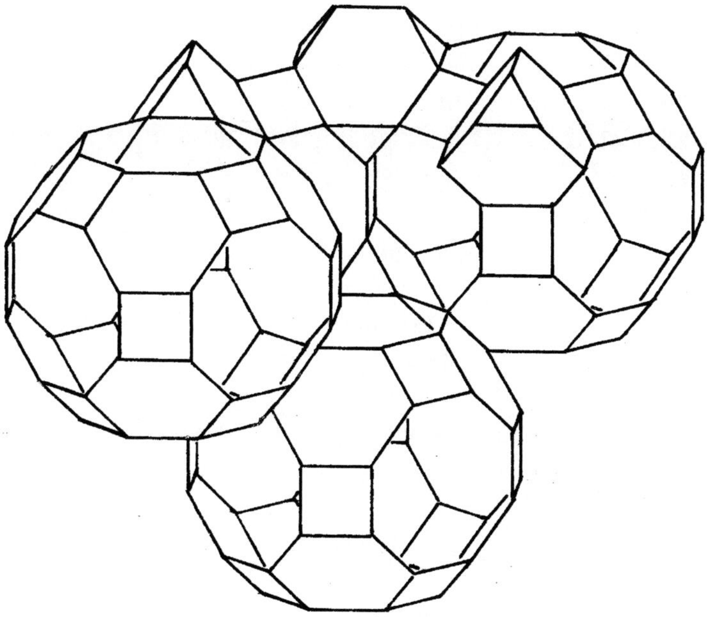

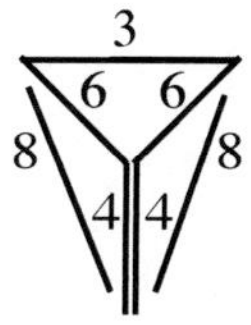

Truncated Cuboctetrahedral Hexaskew [t^3{octet}]

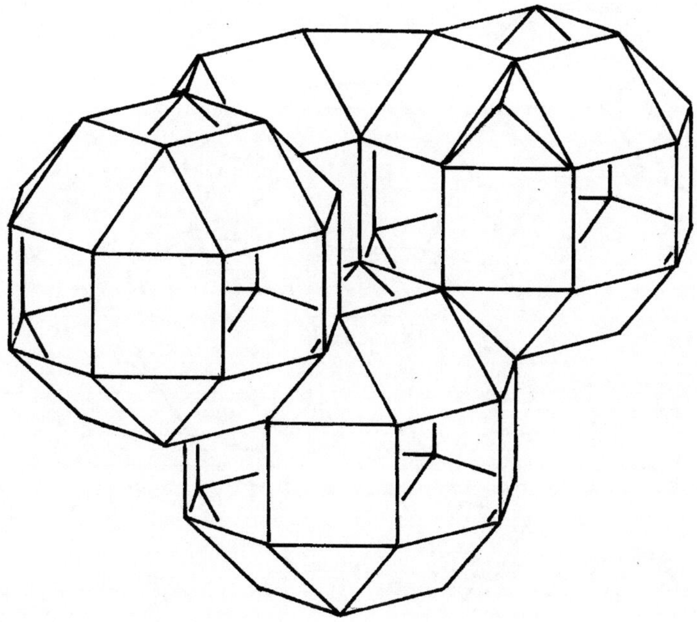

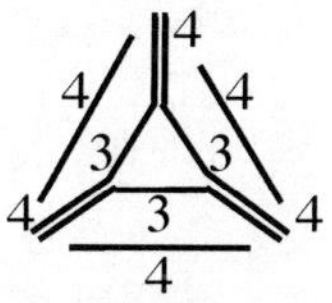

Rhombicuboctetrahedral Hexaskew [t^4{octet}]

Fig.28: Even More Hexaskews

ICOSASKEW POLYHEDRA

Looking in more detail at the truncated cuboctahedral closepack as the central member of this second grid, we can see that the missing hexagons give us 8-connected truncated cuboctahedra and 4-connected truncated octahedra forming one of the spaces.

Similarly the missing squares give us 12-connected truncated cuboctahedra linked to others the same by 2-connected cubes, forming the other space.

This central skew polyhedron is shown below in Figure 29, whilst an intermediate and resultant derived from it are shown opposite in Figure 30, still exhibiting the same connectivity.

Solids	E/V	F/V	S/V
truncated octahedron	5	2 ⬡	1
truncated octahedron		3 □	2
truncated octahedron		(2) ⬡	1
		(1) ‖	
Truncated Octahedral Icosaskew			
cuboctahedron	12	8 □	4
octahedron		4 △	2
		(4) △	
Cuboctahedral Icosaskew		(4) ‖	
trunc cuboctahedron	4	1 ⯃	2
truncated octahedron		1 ⬡	1
cube		2 □	1
		(1) ⬡	
		(1) □	
Truncated Cuboctahedral Icosaskew			

Table 12: Some Icosaskew Polyhedra

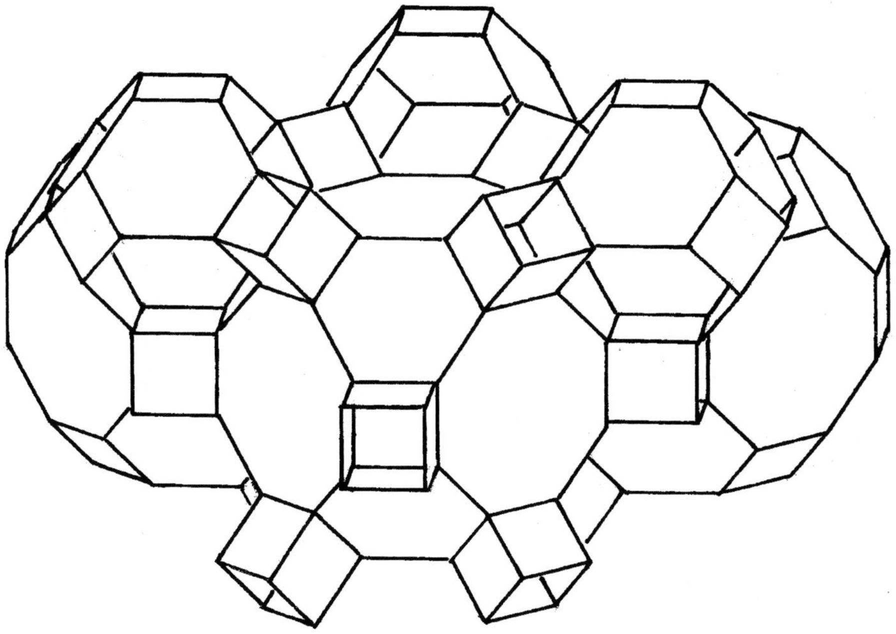

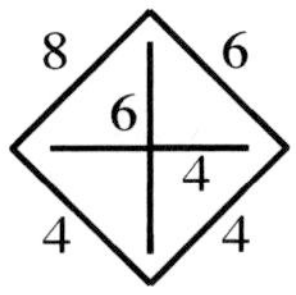

Truncated Cuboctahedral Icosaskew [$tt^2\{4,3,4\}$]

Fig.29: Central Icosaskew Polyhedron

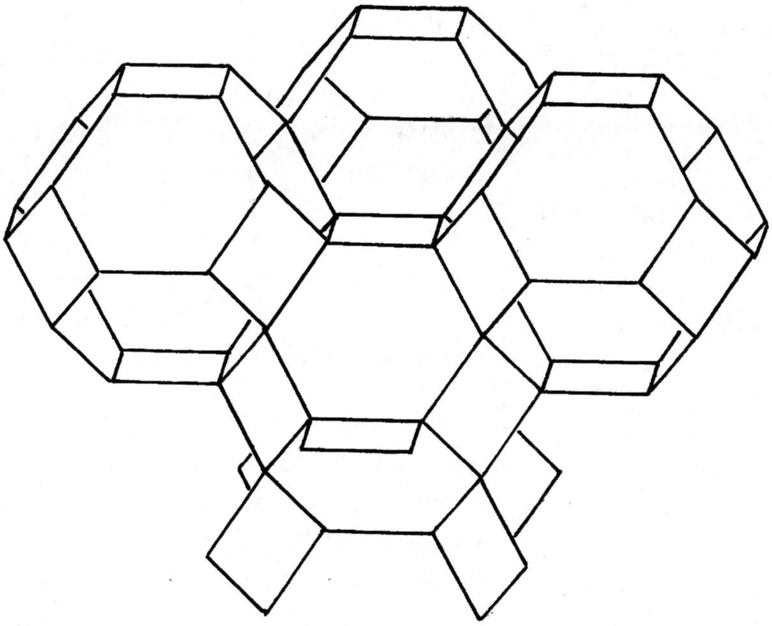

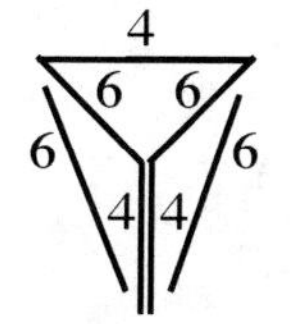

Truncated Octahedral Icosaskew [t^3{4,3,4}] [o{octet}]

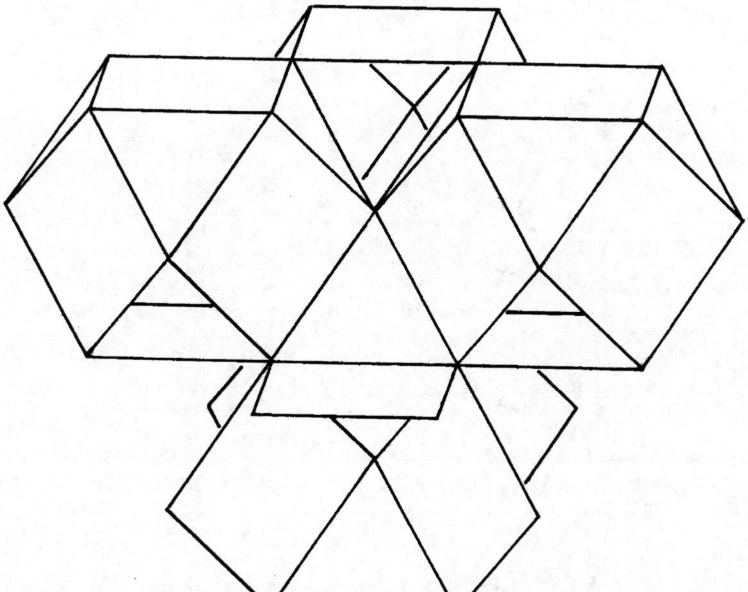

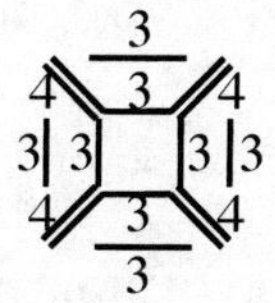

Cuboctahedral Icosaskew [t^2{4,3,4}] [t^2{octet}]

Fig.30: Some Icosaskew Polyhedra

MORE ICOSASKEWS

Three further skew polyhedra in this icosaskew family are shown below and opposite in Figures 31 and 32.

The rhombicuboctahedral icosaskew shown below is an intermediate which can lead to either the resultant cuboctahedral icosaskew shown overleaf or another version of the rhombicuboctetrahedral skew as shown opposite. This is yet another appearance of the t^4\{octet\}, this time retaining its cubes and missing the tetrahedra.

Above this opposite is another intermediate based on the t^3\{octet\}.

Solids	E/V	F/V	S/V
trunc cuboctahedron	5	2 ⯃	2
truncated cube		2 □	1
truncated tetrahedron		1 △	1
		(2) ⬡	
		(1) ‖	
Truncated Cuboctetrahedral Icosaskew			
rhombicuboctahedron	9	9 □	3
cube		(3) △	1
tetrahedron		(3) ‖	1
Rhombicuboctetrahedral Icosaskew			
rhombicuboctahedron	6	5 □	2
cuboctahedron		1 △	1
cube		(2) □	1
		(1) △	
Rhombicuboctahedral Icosaskew			

Table 13: More Icosaskews

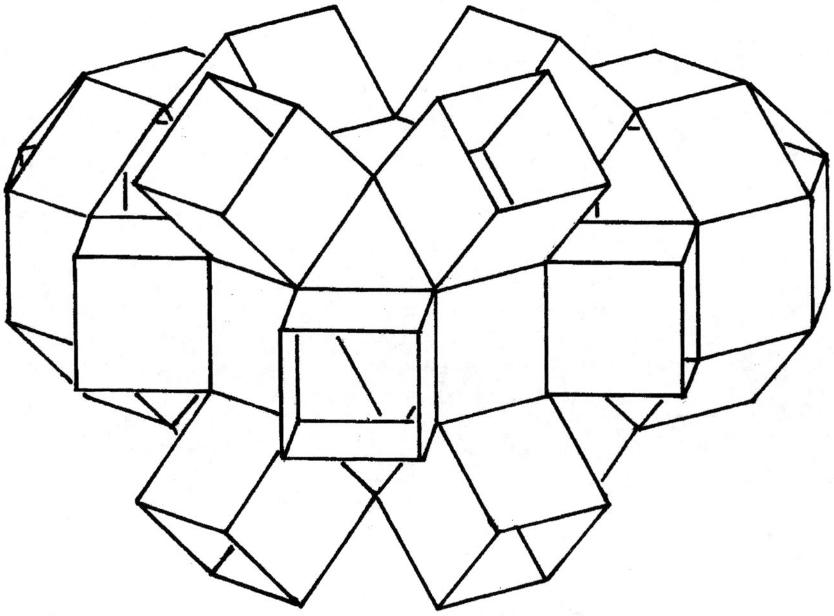

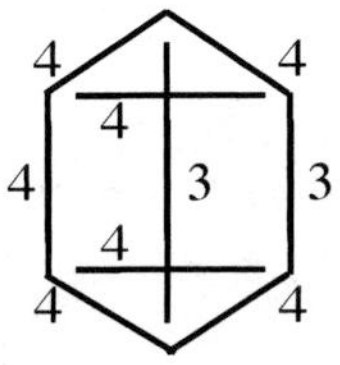

Rhombicuboctahedral Icosaskew $[t^2t^2\{4,3,4\}]$

Fig.31: An Intermediate Icosaskew

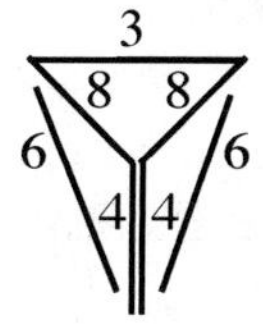

Truncated Cuboctetrahedral Icosaskew $[t^3\{\text{octet}\}]$

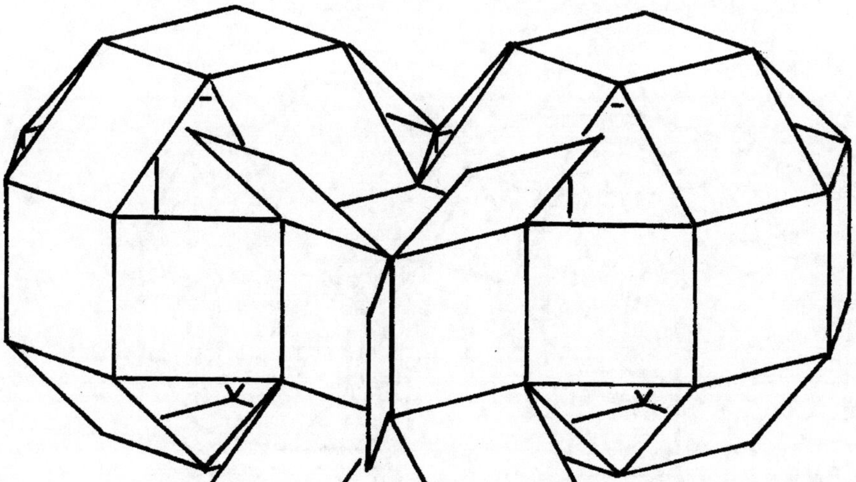

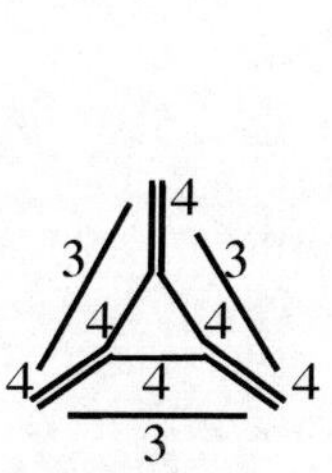

Rhombicuboctetrahedral Icosaskew $[t^4\{\text{octet}\}]$

Fig.32: More Icosaskew Polyhedra

EVEN MORE ICOSASKEWS

Finally we come to the last of this family grid, which comprise three that might form yet another family (see p.41).

The intermediate, named 'cubictetrahedral', has the tetrahedral vertex figure of the truncated cuboctetrahedral icosaskew shown in Figure 32, but with the triangles and squares now missing. The two spaces are comprised of 8-connected truncated cubes linked to 4-connected truncated tetrahedra, along with 12-connected truncated cuboctahedra linked to each other.

The two resultants are shown opposite in Figure 34, one comprising truncated cubes and octahedra, the other truncated octahedra, cuboctahedra and truncated tetrahedra.

Solids	E/V	F/V	S/V
truncated cube	6	4 ⯃	4
octahedron		2 △	1
		(2) △	
		(1) ‖	
Truncated Cubic Icosaskew			
cuboctahedron	6	4 ⬡	1
truncated octahedron		2 □	2
truncated tetrahedron		(2) △	2
		(1) ‖	
Truncated Octetrahedral Icosaskew			
trunc cuboctahedron	4	2 ⯃	2
truncated cube		2 ⬡	1
truncated tetrahedron		(1) □	1
		(1) △	
Truncated Cubictetrahedral Icosaskew			

Table 14: Even More Icosaskews

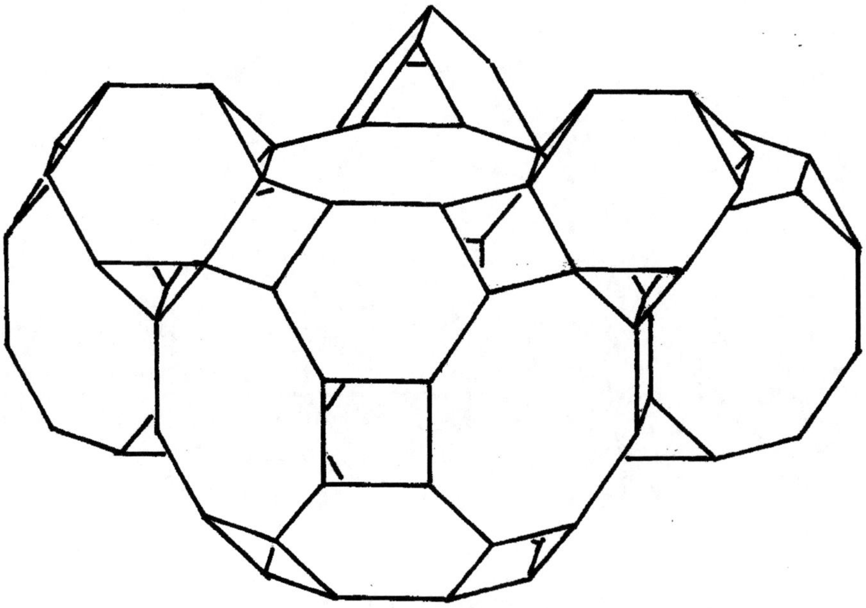

Truncated Cubictetrahedral Icosaskew [t^3 {octet}]

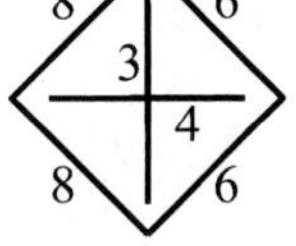

Fig.33: Another Intermediate Icosaskew

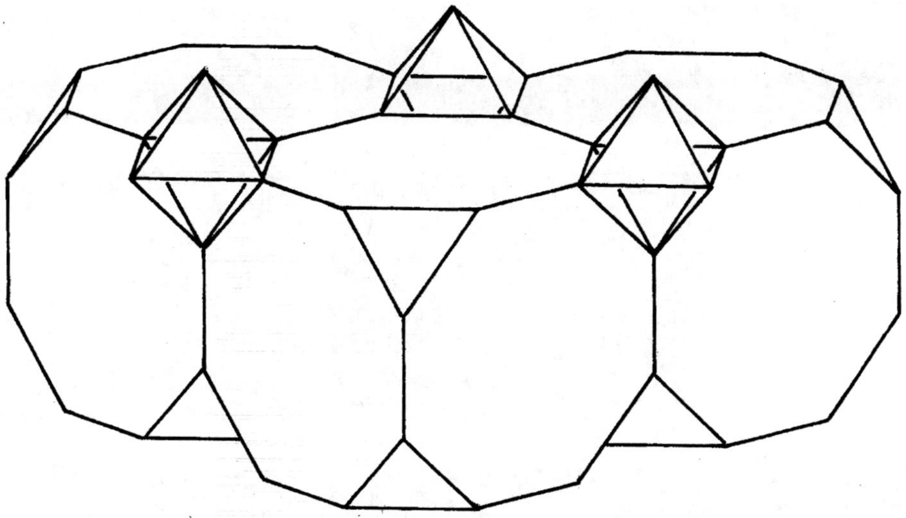

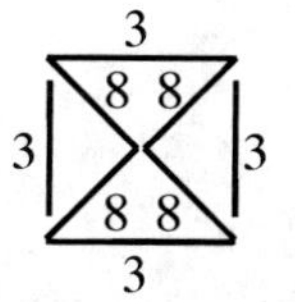

Truncated Cubic Icosaskew [t{4,3,4}]

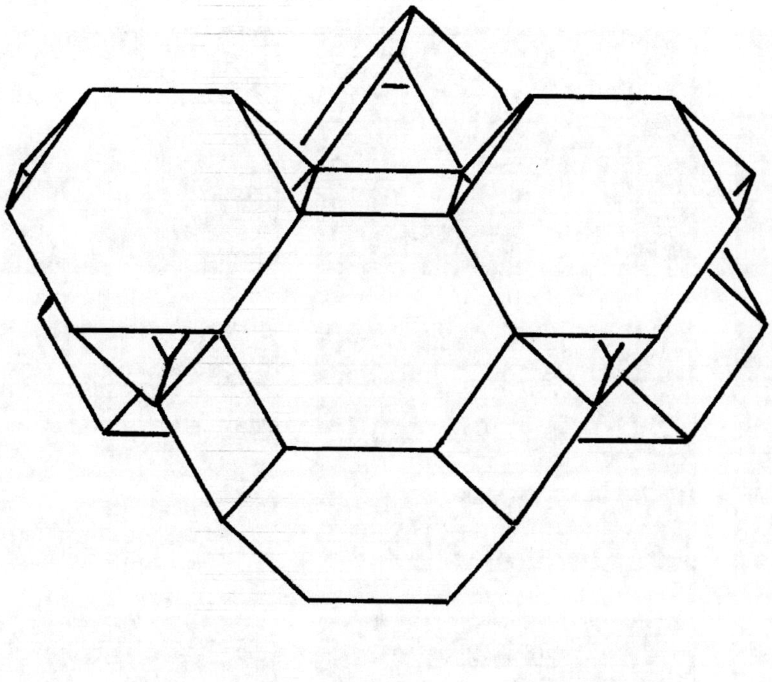

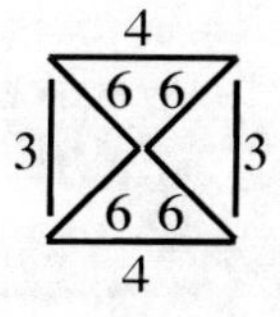

Truncated Octetrahedral Icosaskew [t{octet}]

Fig.34: Resultant Icosaskew Polyhedra

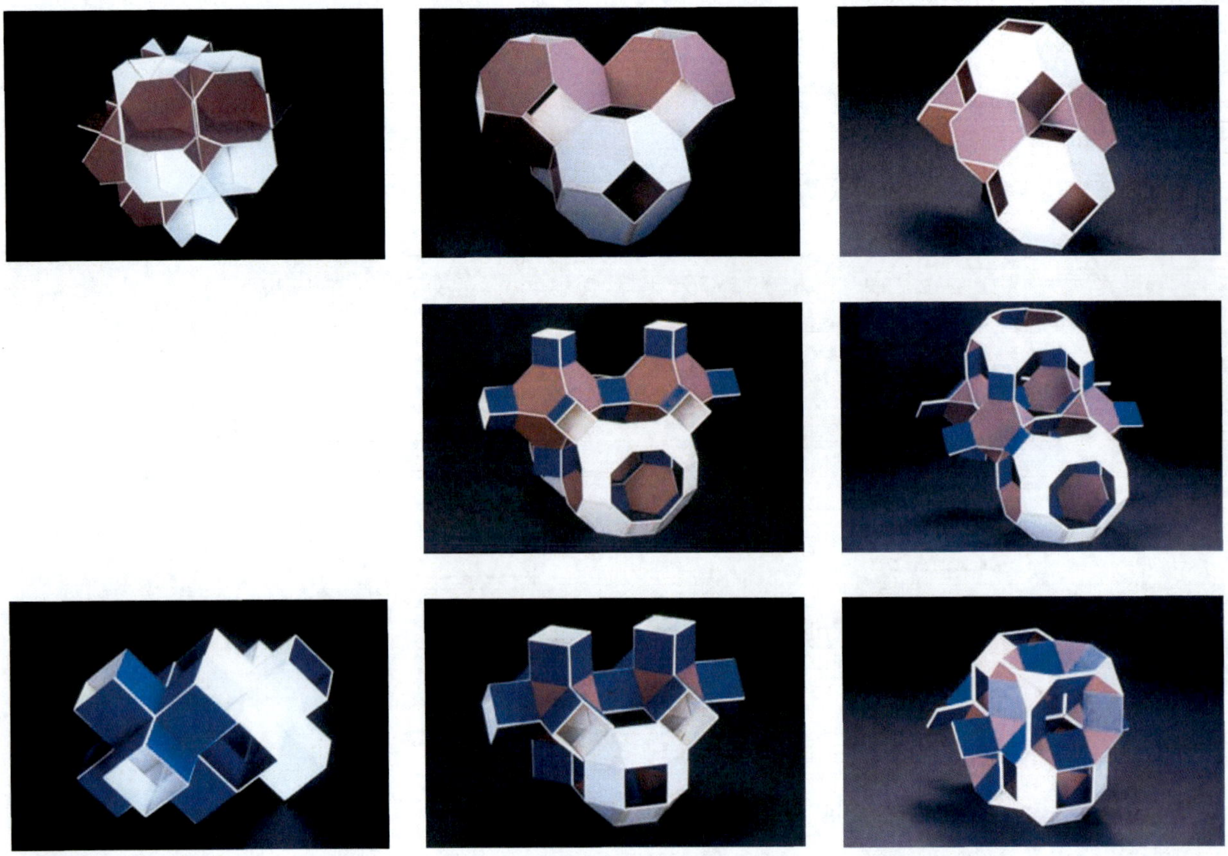

Plate 5: Hexaskew Families [6-6,6-6] and [6-6,2-6]

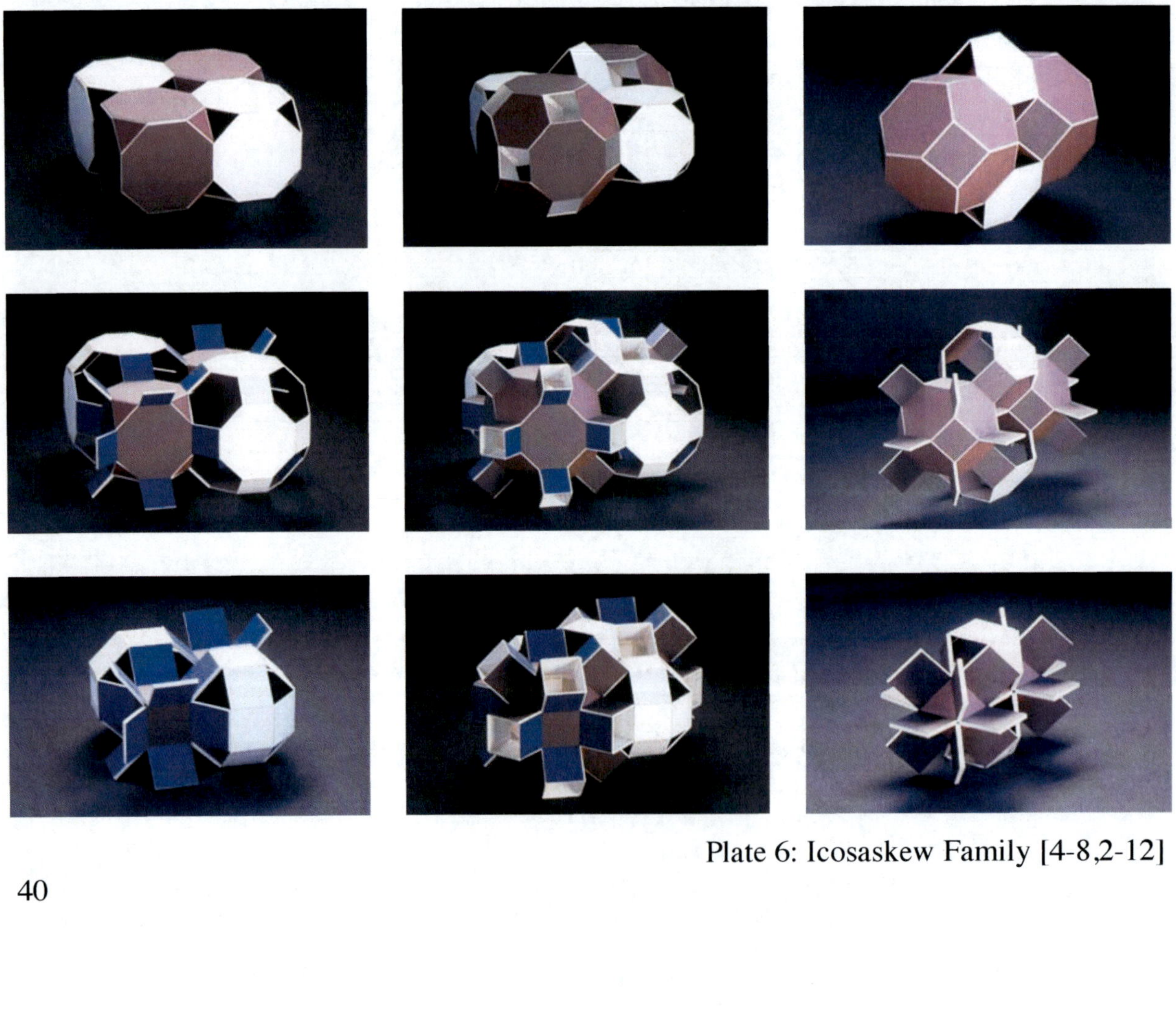

Plate 6: Icosaskew Family [4-8,2-12]

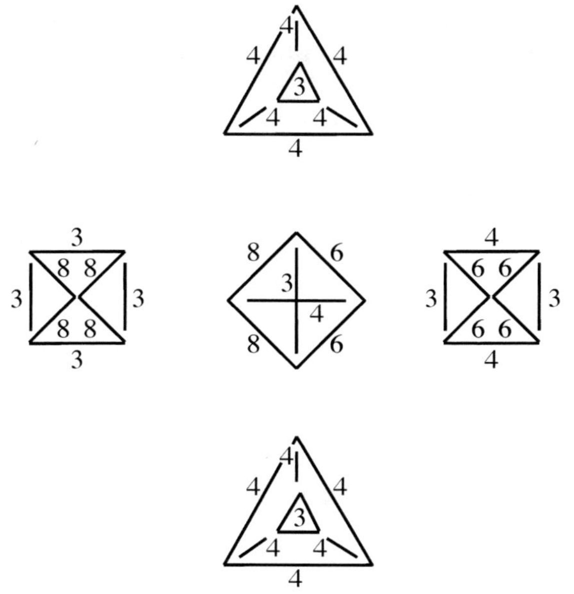

Fig.35: Icosaskew [4-8,12-12] Grid

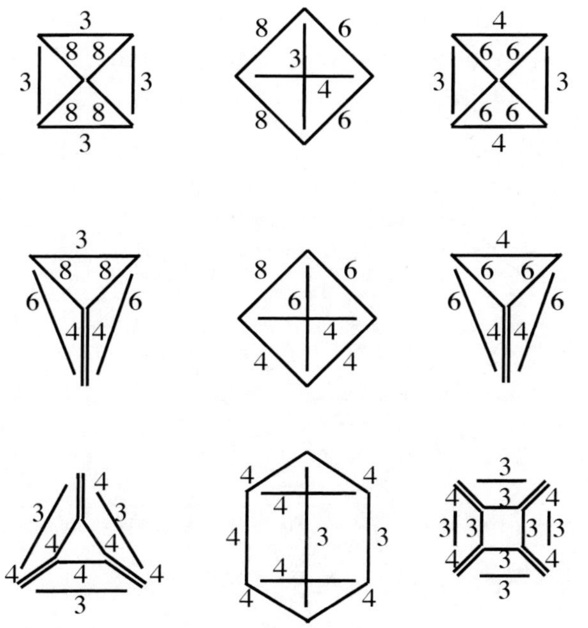

Fig.36: Icosaskew [4-8,2-12] Grid

MORE TRANSFORMATIONS

The grid for these truncated cuboctahedral derived icosaskews is shown below in Figure 36 and Plate 6 opposite shows photographs of models of the members of this family. Again we get appearances of both the t^3 and t^4\{octet\}, and the former with its tetrahedral vertex figure might be thought a suitable subject for further investigation as a central member of yet another family grid.

Such a grid is shown adjoining in Figure 35, but it remains incomplete. The triangular face makes for difficulties as it has an odd number of sides and cannot be considered the result of a truncation. Two intermediate forms are possible, but they only duplicate the resultant forms we saw when the central member was itself an intermediate (see top row of Figure 36 below).

The two other intermediates are the same and comprise a skew polyhedron based around the t^4\{octet\}. One space is formed by 12-connected rhombicuboctahedra linked by their square faces, the other is simply a series of 8-connected cubes and 4-connected tetrahedra, linked only at their vertices and not structurally stable.

With no further tetrahedral vertex figures on which we can operate our transformations, we seem to be at the end of our description of the skew polyhedra. We are exactly that in one respect as this method will produce no further examples that fill all of three dimensions. A short summary would now seem to be in order, linking these skew polyhedra to the possible sphere packings.

There is however more to look at, as we will then go back to look at our original concept of notional truncation, which will open up a whole new vista for exploration.

SPHERE PACKINGS

We have seen in our explorations of the skew polyhedra that the surfaces produced, polygonal tilings in three dimensions, all fit in and around the various available closepacks of regular and semi-regular polyhedra.

The various connectivities we have seen in the cells of these can be explained by treating the constituent polyhedra as simply spheres in some form of regular packing touching other spheres near them. The connections then simply reduce to three dimensional lattices, the edges of which connect the centres of adjacent touching spheres.

In crystallography the simplest such assembly of spheres is that of 'simple cubic packing' (scp), each sphere touching six others at the vertices of an octahedron. The spheres are as if arranged inside a box and the basic cell has eight touching spheres. The assembly has a density of spheres to total space of 0.5236 and the corresponding lattice forms a cubic 6-connected grid with everything at right angles.

Spreading these eight spheres slightly further apart to allow the insertion of an extra one at the centre of each basic cell we get 'body centred cubic' packing (bcc). Here each sphere touches eight others at the vertices of a cube. The density now increases to 0.6801 and the corresponding lattice follows the edges of a rhombic dodecahedral closepack with an extra node at the centre of each rhombic dodecahedron.

Instead of inserting one extra sphere in the centre of our original simple cubic cell, we can spread the eight spheres even further apart and insert six extra spheres at the centres of each face. This gives rise to face centred cubic packing (fcc), where surprisingly each sphere actually touches 12 others at the vertices of a cuboctahedron. This is the classic packing obtained by greengrocers with their oranges and has the maximum possible density of 0.7405.

The 12-connected lattice for this packing is that of the octet truss, the edges following those of closepacked octahedra and tetrahedra.

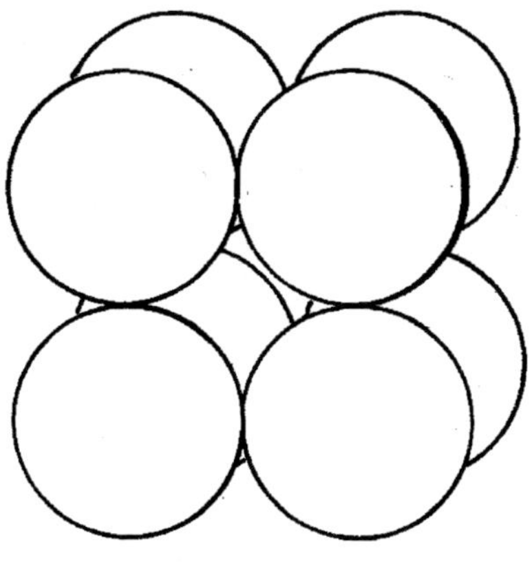

Fig.37: Simple Cubic Packing (scp)

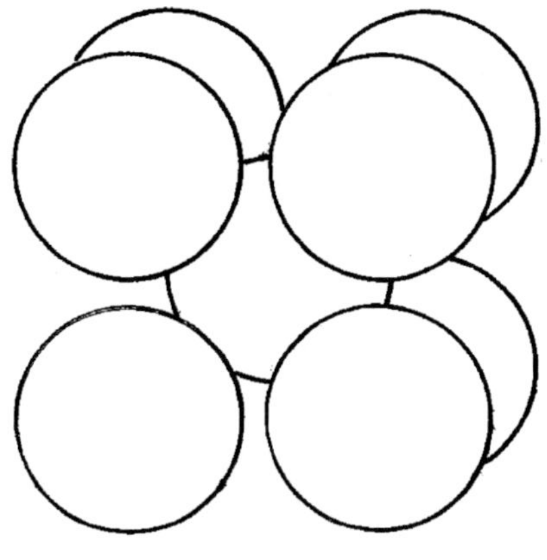

Fig.38: Body Centred Cubic Packing (bcc)

TETRASKEWS AND HEXASKEWS

If we now return to body centred cubic packing with its central sphere and eight connections, we can omit four of these at alternate corners of the cube to obtain 'tetrahedral' packing (tet), where each sphere touches only four others at the vertices of a tetrahedron. This has a reduced density of 0.3401 and its corresponding 4-connected lattice, as found in the structure of diamond, follows some of the edges of a rhombic docecahedral closepack.

With this lattice, the nodes at the centres of the missed out spheres form an identical set of 4-connected points that weave in and around the ones we have retained. We thus get a dual pair of tetrahedral lattices that between them take in all the points of our original body centred cubic array.

These paired lattices form the underlying connection framework of the 4-connected family of tetraskew polyhedra, which can thus be seen as polygonal tilings of the surface separating the two lattices. This if smoothed out to remove the polygonal sharp corners is the D-type minimal surface.

It will be recalled that the 4-connected truncated octahedral tetraskew is the central member of the tetraskew family. The realignment of its vertex figure, missing out different opposing polygons, leads to the 6-connected truncated octahedral hexaskew.

This is also based around the truncated octahedral closepack, but the cells here are connected by their (missing) square faces along the lines of a 6-connected simple cubic lattice. Another set of 6-connected points is formed by the missed out points at the centre of each cubic cell, so again we get a dual pair of lattices, this time both simple cubic, which taken together give us back the original body centred cubic array.

This different pair of lattices thus forms the underlying connection framework of the 6-connected family of hexaskew polyhedra.

These can also be seen as polygonal tilings of the surface separating the two lattices, which if smoothed out to remove the polygonal sharp corners becomes what is known as the P-type minimal surface.

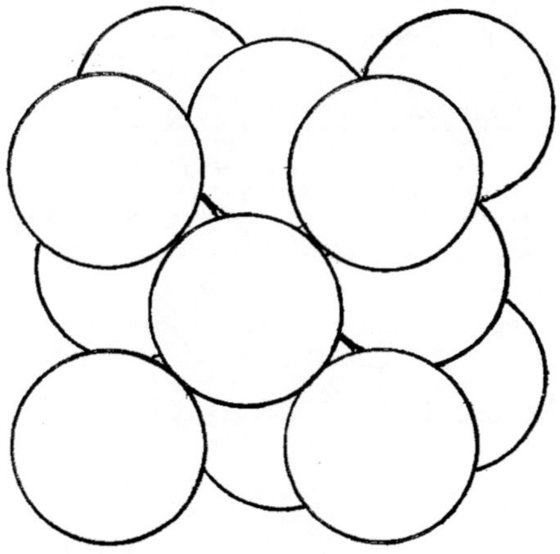

Fig.39: Face Centred Cubic Packing (fcc)

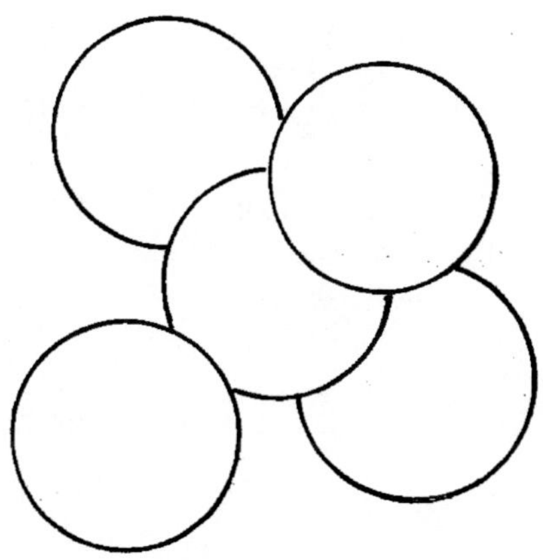

Fig.40: Tetrahedral Packing (tet)

OCTASKEWS AND DODECASKEWS

The truncated octahedral hexaskew central to this hexaskew family's grid, also occurs as a 'parent' skew polyhedron in the family with the truncated cuboctaprism hexaskew as its central member. Here as well as the 6-connected truncated cuboctahedra, there are 2-connected octaprisms present.

If we incorporate these octaprisms into our sphere and lattice description we effectively get two interpenetrating simple cubic lattices with extra 2-connected nodes central to their edges, each of which might be considered as edge centred cubic packing. The two sets of spheres comprising these two lattices taken in their totality deliver us back to full simple cubic packing again.

The octaskews produced by the second version of this vertex have as their central member the same truncated cuboctaprism closepack, but now the truncated cuboctahedra are 8-connected via missing hexagon faces, essentially a body centred cubic array of polyhedra, whilst the other space is formed of 4-connected octaprisms, each one set perpendicular to the next.

The nature of this second space is well illustrated by the rhombic {4,8} octaskew that encloses it, shown in Plate 3. The 4-connected lattice following the centres of this space taken together with that of the body centred cubic truncated cuboctahedra brings us back to the full simple cubic packing we started with.

Simple cubic is understandably the resultant arrangement when we look at the family produced by the third orientation of this vertex of the truncated cuboctaprism closepack. The dodecaskew version of this comprises truncated cuboctahedra in a body centred cubic arrangement, divided into two sets of simple cubic each with octaprisms interposed at the centres of the cell faces. The result is two sets of interlocking face centred cubic, each 4- and 12-connected.

The remaining family of hexaskews is formed by taking an intermediate from the the truncated cuboctaprism hexaskew family, the truncated cuboctahedral hexaskew, as the central member of its own family grid.

Centre vertex:	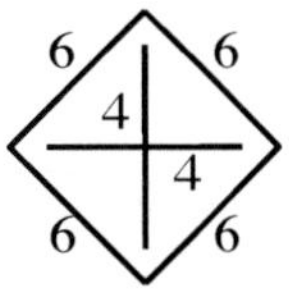	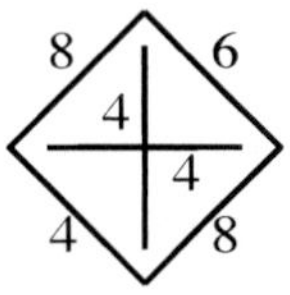	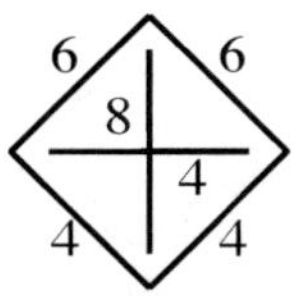	
Truncated:	Octahedral	Cuboctaprism	Cuboctahedral	Cuboctetrahedral
Tetraskews	[4-4,4-4]			
Hexaskews	[6-6,6-6]	[2-6,2-6]	[6-6,2-6]	
Octaskews		[4-4,8-8]		
Dodecaskews		[4-12,4-12]		
Icosaskews			[4-8,2-12]	[4-8,12-12]

Table 15: The Families of Skew Polyhedra

AND ICOSASKEWS

Here the truncated cuboctahedra are all 6-connected in simple cubic packing, but the similarly 6-connected truncated octahedra have extra 2-connected cubes between them, thus forming an edge centred cubic lattice. The overall result here is a cubic packing expanded by both body and edge centred extra spheres. Our final family of icosaskews is formed by taking this last truncated cuboctahedral vertex in a different orientation, missing out a hexagon and a square instead of an octagon and square.

The resultant icosaskews have one space both 4- and 8-connected forming a rhombic dodecahedral lattice, whilst the other space comprises 12-connected polyhedra arranged around an octet truss lattice that is effectively edge centred by the insertion of extra 2-connected nodes. These last occur at the centres of the rhombs forming the other lattice. The totality of these is again understandably an array of polyhedra in a cubic packing expanded by both body and edge centred extra spheres.

The one remaining family grid was incomplete, formed by taking the truncated cuboctetrahedral closepack as its central member. Members of this family again have one space 4- and 8-connected, whilst the other is simply 12-connected, without the extra nodes seen in the last family. The overall packing here is body centred cubic.

Overall then we have five distinct sets of connections, as set out in Table 15 opposite and shown in Plate 7 adjoining. It will be seen that the names 'tetra,, 'hexa', 'octa' and 'dodeca' relate directly to the numbers of connections encountered, whilst 'icosa' fits as 8+12. This has no connection to the same prefixes applied to the polyhedra with the symmetries {3,3}, {3,4} and {3,5}.

Tetraskews
[4-4,4-4]

Hexaskews
[6-6,6-6]

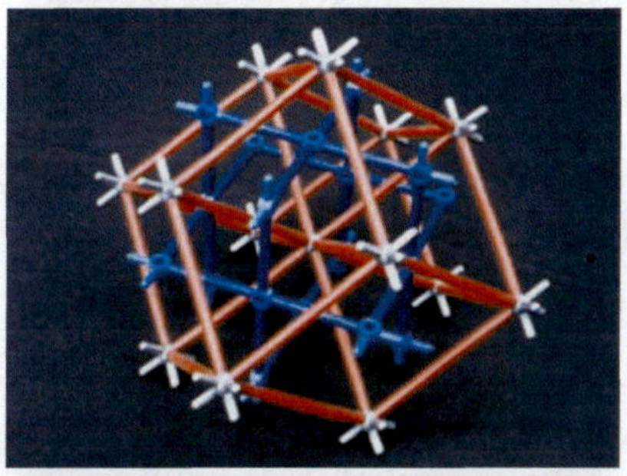

Octaskews
[4-4,8-8]

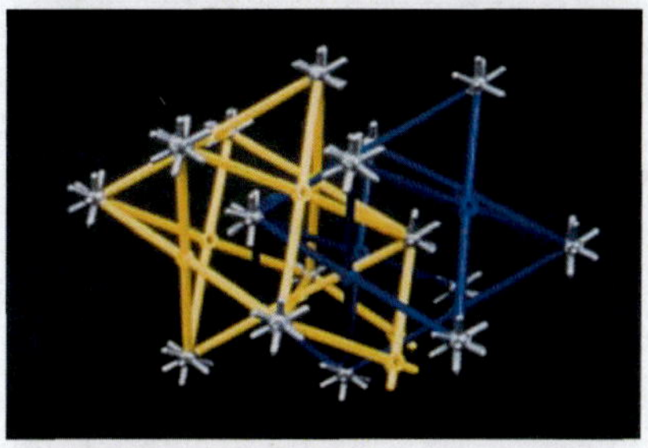

Dodecaskews
[4-12,4-12]

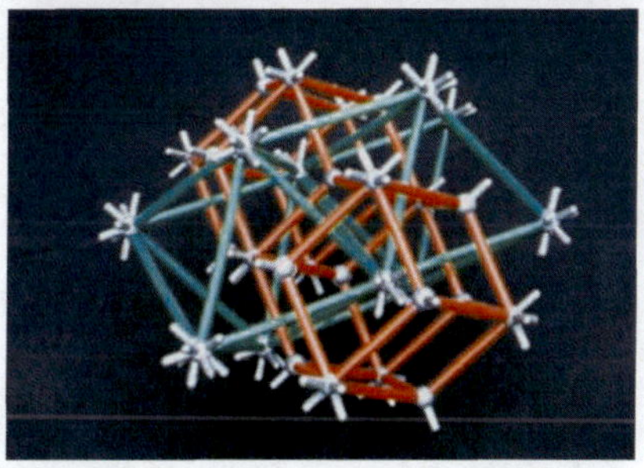

Icosaskews
[4-8,12-12]

Plate 7: The Five 3-D Lattices

DEGENERATE SKEW POLYHEDRA

Although we seem to have exhausted the possibilities using vertex transformation on tetrahedral vertex figures from the closepacks, we can indeed do more by considering once again the 'notional truncation' we first noticed with the skews.

The 4-connected {6,6} and 6-connected {4,6} families are fairly intertwined using this, but the 4- and 8-connected {8,8} family does not seem to have a counterpart.

Two plus two octagons at a vertex would require there to be an {8,4}, which would have a dual {4,8} and a central grid member with one octagon, one square and two further squares at each vertex.

Figures 41 below and 42 opposite show how these combinations can be achieved as parts of a 4-connected family of degenerate tetraskew polyhedra.

Solids	E/V	F/V	S/V
tetratile	8	8 □	1
cube		(4) □	4
		(4) ‖	
{4,8 |4}			
Tetraprism Tetraskew			
truncated tetratile	4	4 ⯃	2
		(1) □	
		(1) ‖	
{8,4 |4}			
Truncated Tetratile Tetraskew			
truncated tetratile	4	1 ⯃	1
octaprism		3 □	2
cube		(1) ⯃	1
		(1) □	
Truncated Tetratetraprism Tetraskew			

Table 16: Degenerate Tetraskews

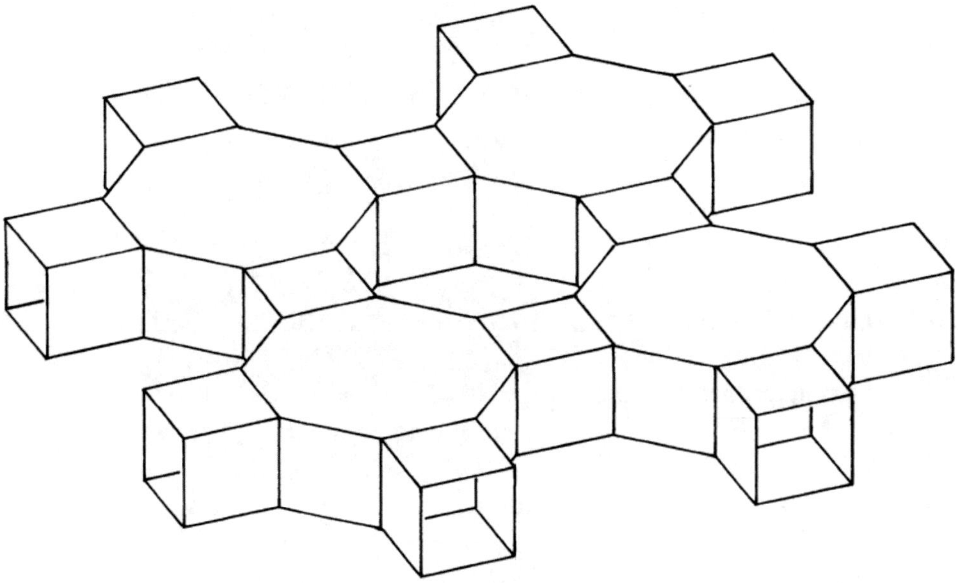

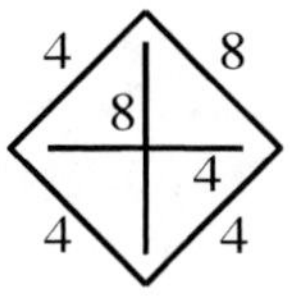

Truncated Tetratetraprism Tetraskew [o{4,4,2}]

Fig.41: Central Degenerate Tetraskew

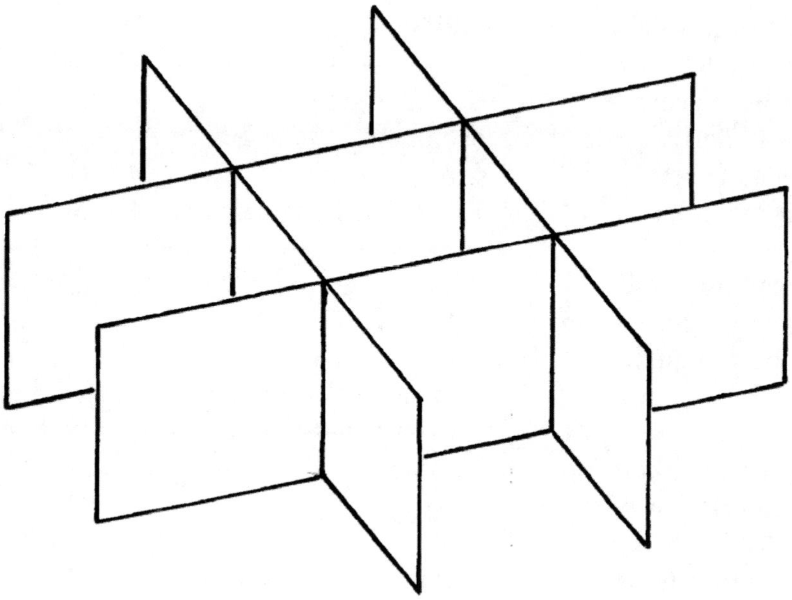

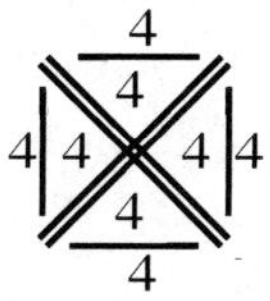

{4,8 |4}
Tetraprism Tetraskew $[o^3\{4,4,2\}]$

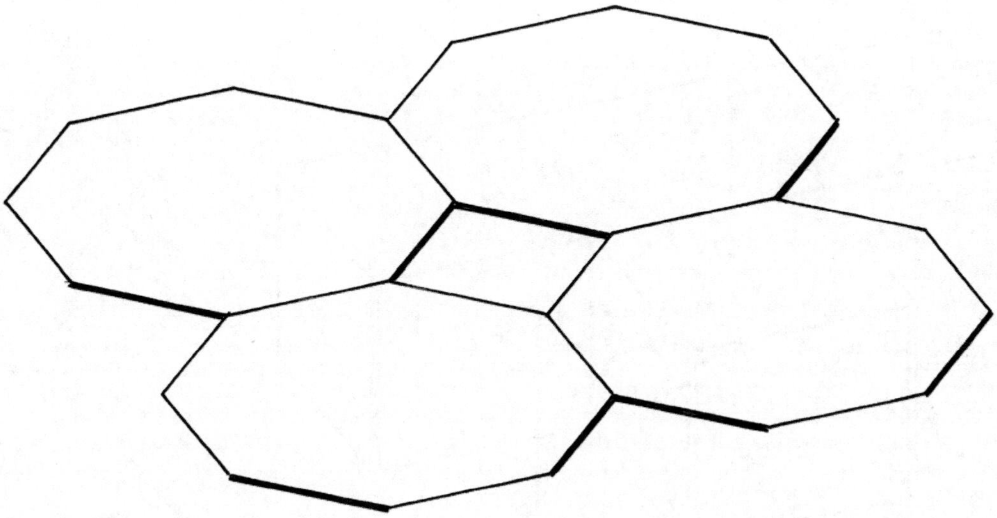

{8,4 |4}
Truncated Tetratile Tetraskew $[t^3\{4,4,2\}]$

Fig.42: Parent Degenerate Tetraskews

MORE DEGENERATE SKEWS

Taking the {4,8} and {8,4} of Figure 42 as parent skew polyhedra and the tetrahedral vertex of that in Figure 41 as the central member of their grid, we can derive the rest of the family by our tried and tested technique of vertex transformations.

Two intermediate forms leading to the parents are shown opposite in Figure 44 and a resultant of these other than a parent in Figure 43 below.

As with the parents these are all either entirely two dimensional and flat, or on occasion only one square deep, where a digon has been truncated, venturing in a very limited way into a third dimension.

Solids	E/V	F/V	S/V
truncated tetratile	4	2 ⯃	1
octaprism		2 □	2
cube		(2) □	1
Truncated Tetraprism Tetratetraskew			
truncated tetratile	5	5 □	1
octaprism		(2) ⯃	2
cube		(1) ‖	1
Truncated Tetraprism Tetraskew			
tetratile	6	6 □	1
cube		(2) □	4
		(1) ‖	
Tetratetraprism Tetraskew			

Table 17: More Degenerate Tetraskews

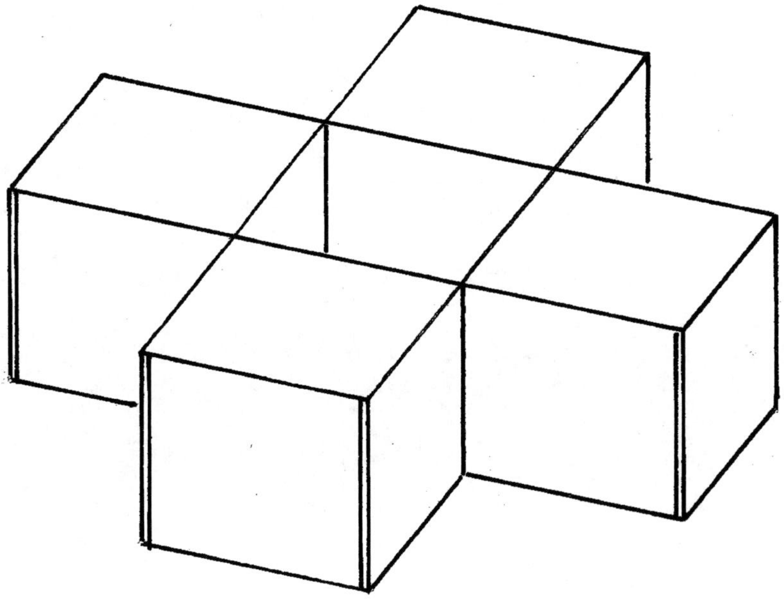

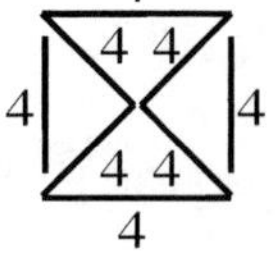

Tetratetraprism Tetraskew $[t^2t^2\{2,4,4\}]$

Fig.43: Resultant Degenerate Tetraskew

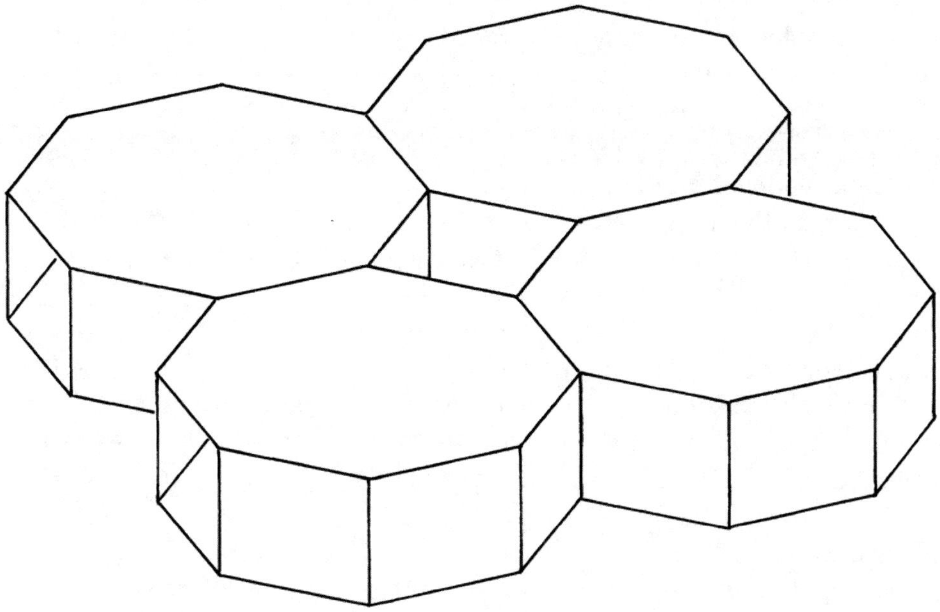

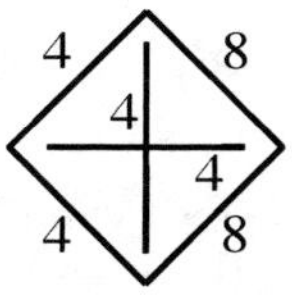

Truncated Tetraprism Tetratetraskew [tt^2{2,4,4}]

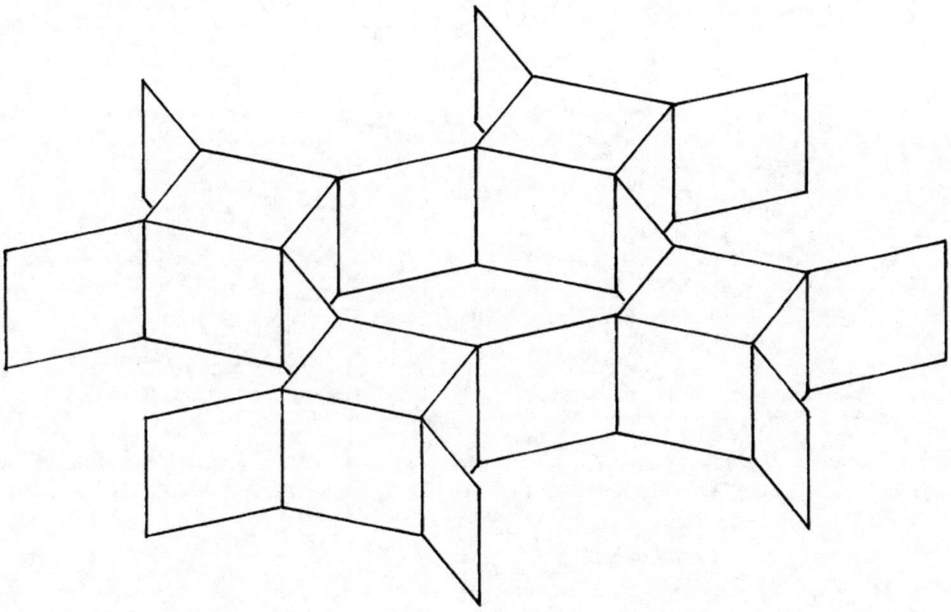

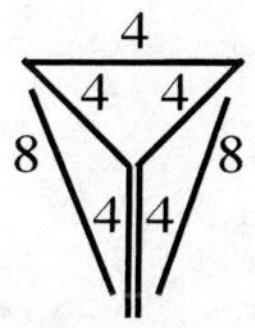

Truncated Tetraprism Tetraskew [o^2{2,4,4}]

Fig.44: Intermediate Degenerate Tetraskews

FURTHER DEGENERATE SKEWS

We do in fact have a complete family here with two parents with missing square faces, {8,4 |4} and {4,8 |4} in Coxeter's notation.

The remaining two intermediates and one resultant skew polyhedra in this family are shown opposite and below in Figures 46 and 45 respectively.

They are all 4-connected along the lines of a square tiling {4,4} in two dimensions, rather than a diamond lattice in three dimensions as were the {6,6 |3} family. They all go to infinity in two dimensions, although sometimes the connections squeeze through missing digons between back to back polygons.

Degenerate skews will now be found linked to the other two regular tilings, that of the hexagon {6,3} and triangle {3,6}.

Solids	E/V	F/V	S/V
truncated tetratile	4	2 ⯃	2
		2 □	
		(1) ⯃	
		(1) ∥	
Truncated Tetratetratile Tetraskew			
tetratile	5	5 □	1
cube		(3) □	4
Rhombitetratetraprism Tetraskew			
tetratile	6	6 □	2
		(1) □	
		(2) ∥	
Rhombitetratetratile Tetraskew			

Table 18: Further Degenerate Tetraskews

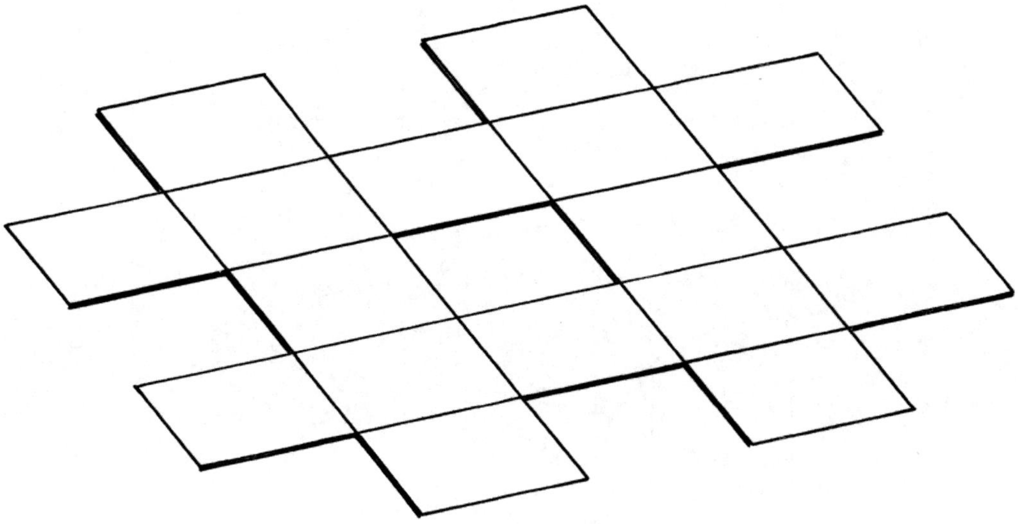

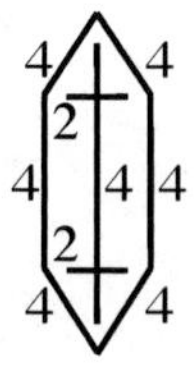

Rhombitetratetratile Tetraskew $[t^2t^2\{4,4,2\}]$

Fig.45: Another Resultant Degenerate

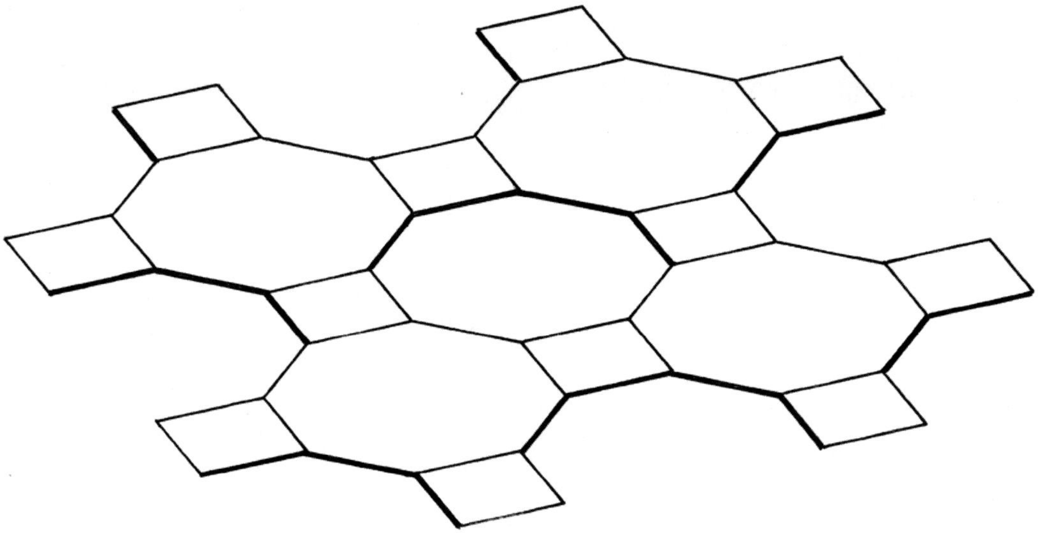

Truncated Tetratetratile Tetraskew $[tt^2\{4,4,2\}]$

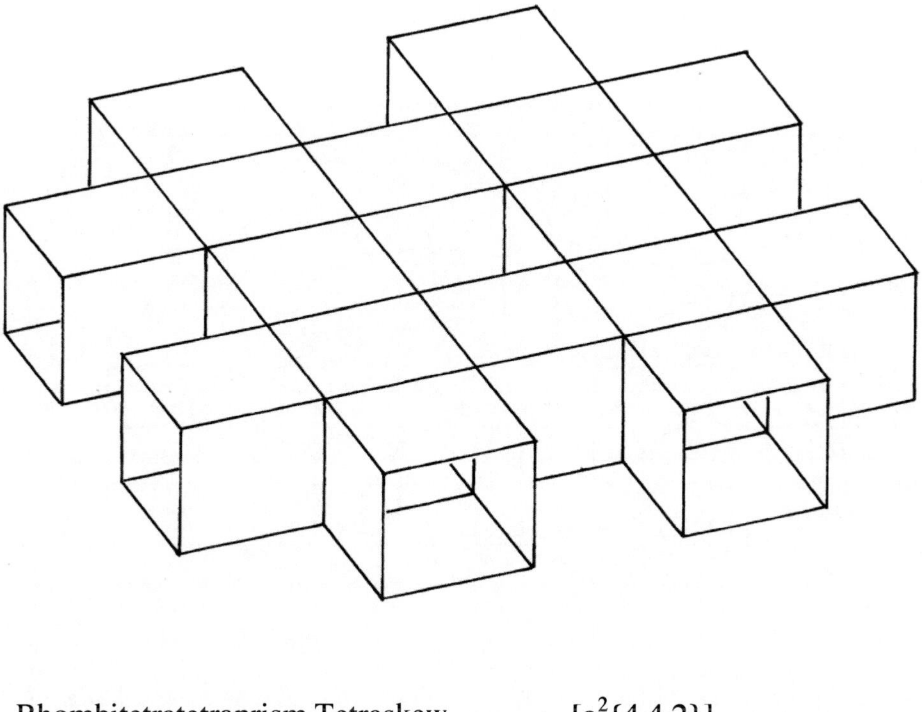

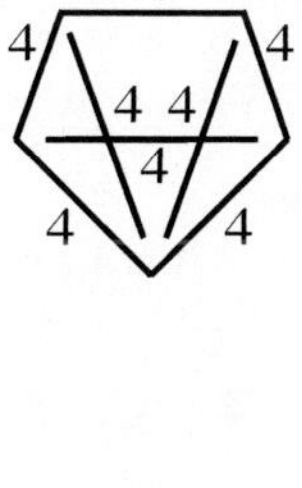

Rhombitetratetraprism Tetraskew $[o^2\{4,4,2\}]$

Fig.46: More Intermediate Tetraskews

ANOTHER DEGENERATE FAMILY

If we now go back to our original {6,4 |4} vertex, we find it can be bent back in a similar way to that of the {8,4 |4} to produce another regular form. Effectively {6,4 |6}, this is a 3-connected form of the hexagon lattice with some faces missing.

Not unsurprisingly it has a dual {4,6 |6} formed by bending back the {4,6 |4} vertex until the squares are back to back enclosing another 3-connected space, mimicking the hexagonal cells of a honeycomb.

Looking for the central member of such a family, it should have one hexagon, one square and two further squares at each vertex, again a bent back version of the {6,4 |4} central member.

Figures 47 below and 48 opposite show these combinations as parts of a further family of degenerate skew polyhedra.

Solids	E/V	F/V	S/V
hexatile	6	6 □	1
hexaprism		(3) ⬡	3
		(3) ‖	
{4,6 |6}			
Hexaprism Triskew			
hexatile	4	4 ⬡	2
		(1) ⬡	
		(1) ‖	
{6,4 |6}			
Truncated Tritile Triskew			
truncated trihexatile	4	1 ⬡	1
dodecaprism		3 □	1
hexaprism		(1) ○	1
cube		(1) □	1
Truncated Trihexaprism Triskew			

Table 19: Degenerate Triskews

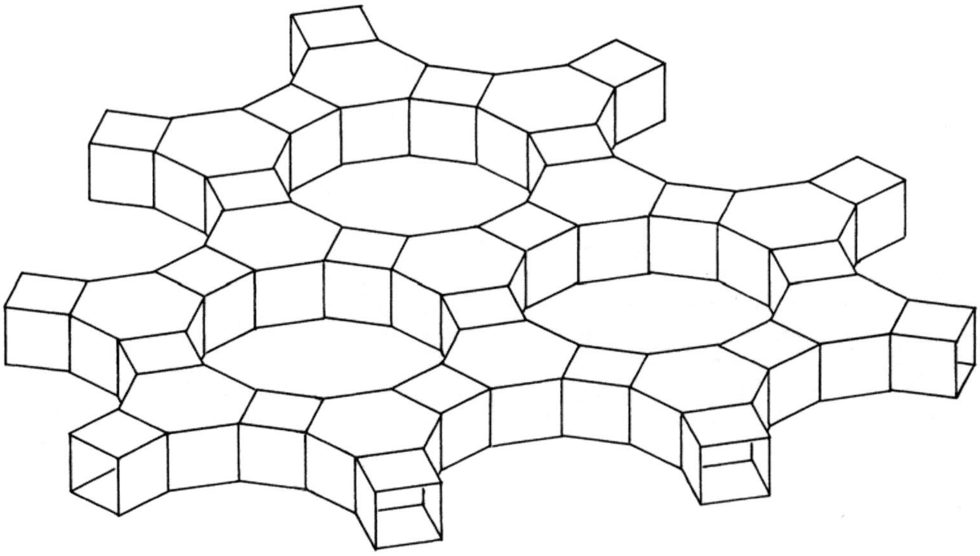

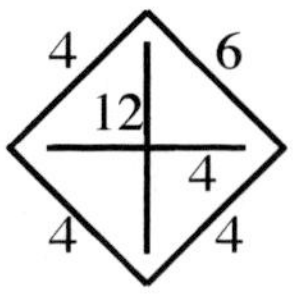

Truncated Trihexaprism Triskew [o{6,3,2}]

Fig.47: Central Degenerate Triskew

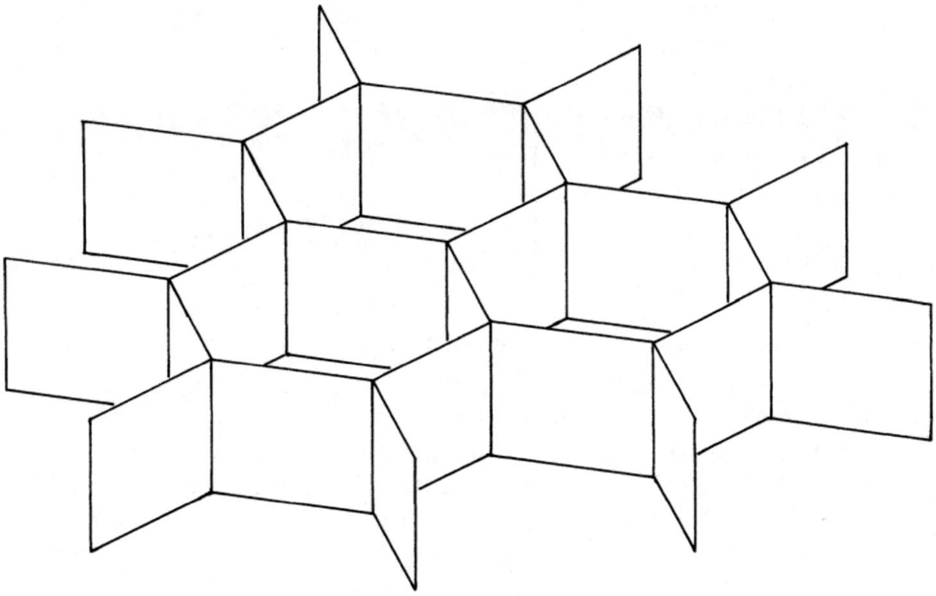

{4,6 |6}
Hexaprism Triskew $[o^3\{6,3,2\}]$

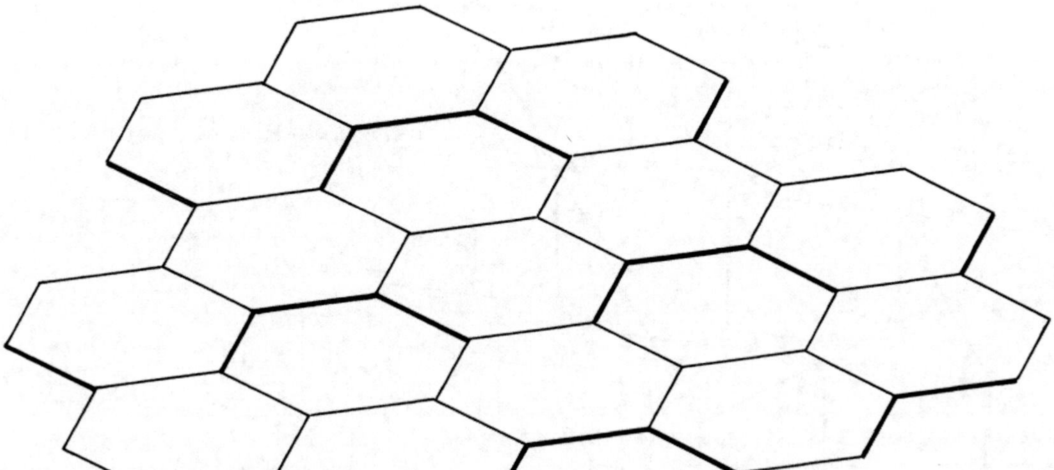

{6,4 |6}
Truncated Tritile Triskew $[t^3\{6,3,2\}]$

Fig.48: Parent Degenerate Triskews

MORE DEGENERATE TRISKEWS

Whilst the {4,6 |4} family only produced two intermediates and one resultant skew polyhedron other than the parents, the {4,6 |6} family we are now looking at can double those numbers. This is because the central member's vertex figure is less symmetrical missing a square and a dodecagon rather than two equal octagons.

Effectively what happens here is that the vertices of the original {4,6 |4} family can be folded back in two different ways to produce flattened vertex figures with different arrangements of the same polygons.

Opposite and below in Figures 50 and 49 are one pair of these intermediates and one resultant.

Solids	E/V	F/V	S/V
hexatile	4	2 ⬡	1
hexaprism		2 □	3
		(1) ⬡	
		(1) □	
Truncated Triprism Triskew			
truncated hexatile	5	4 □	1
dodecaprism		1 △	2
triprism		(2) ◯	1
		(1) ∥	
Truncated Hexaprism Triskew			
trihexatile	6	4 □	1
hexaprism		2 △	2
triprism		(2) ⬡	2
		(1) ∥	
Trihexaprism Triskew			

Table 20: More Degenerate Triskews

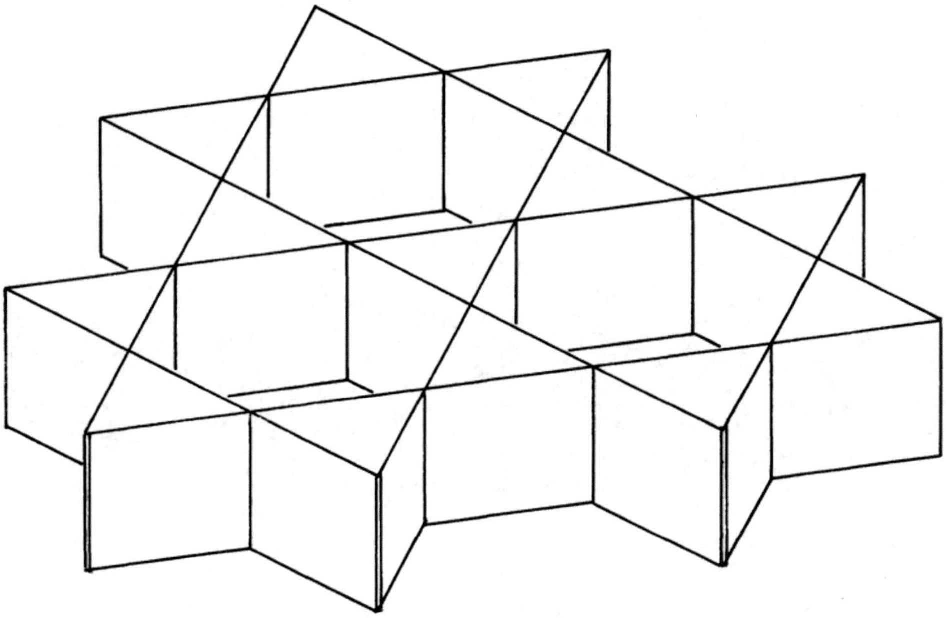

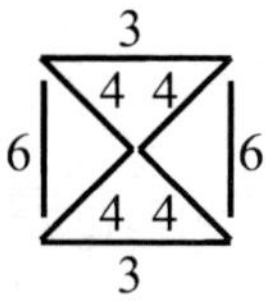

Trihexaprism Triskew [t^2t^2{2,3,6}]

Fig.49: Resultant Degenerate Triskew

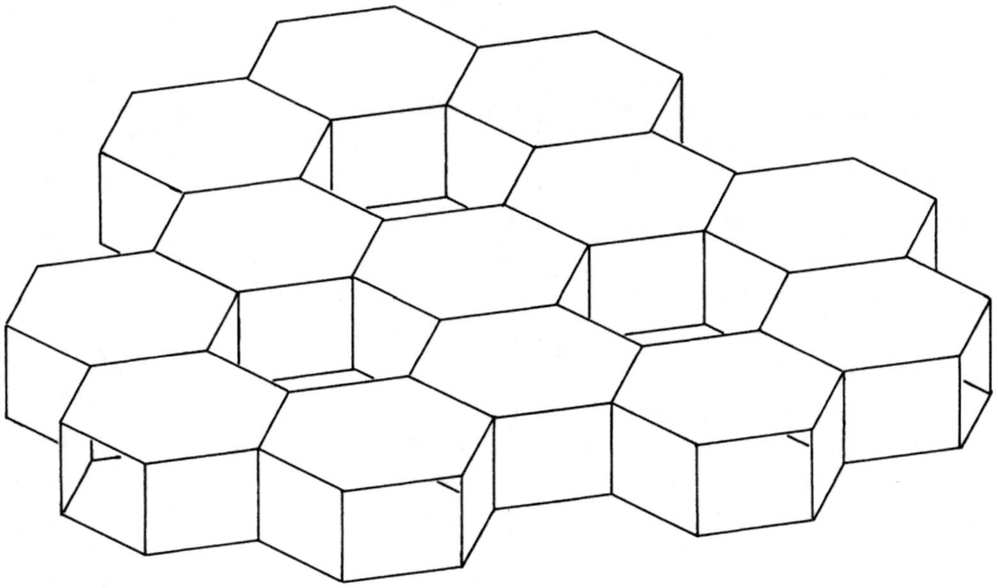

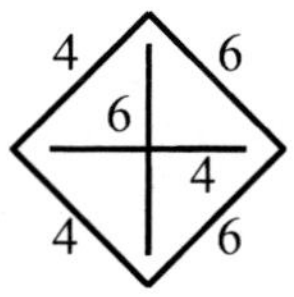

Truncated Triprism Triskew [$tt^2\{2,3,6\}$]

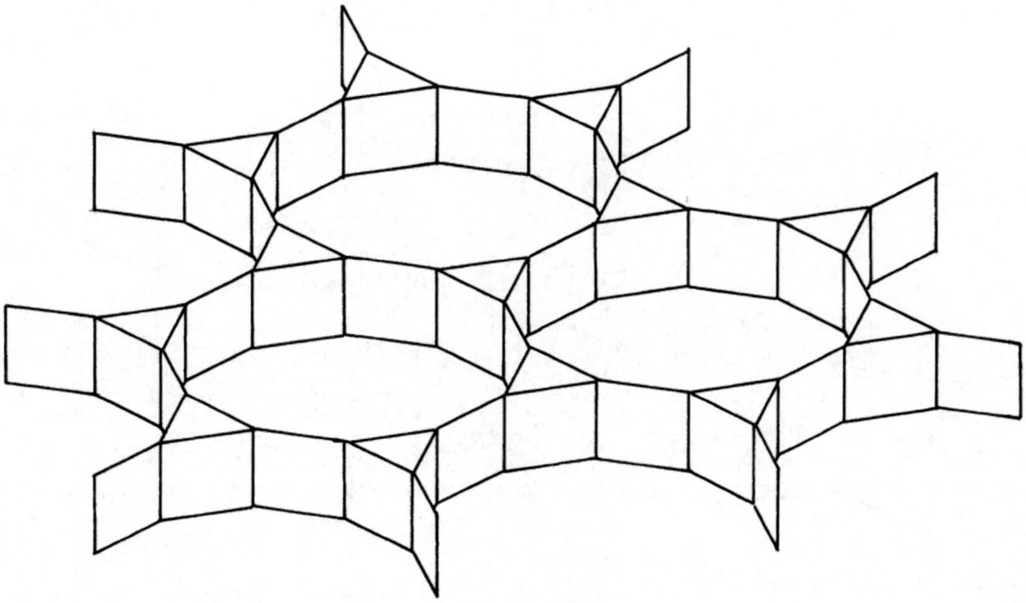

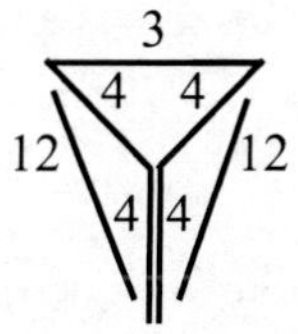

Truncated Hexaprism Triskew [$o^2\{2,3,6\}$]

Fig.50: Intermediate Degenerate Triskews

EVEN MORE TRISKEWS

Following through to find the other members of this family grid there are a second pair of intermediates and another resultant as shown opposite and below in Figures 52 and 51.

Comparing these one to one with the first set of such in Figures 50 and 49, they can be seen to exhibit exactly the same polygons at each vertex, but arranged differently.

With this grid now filled in, it will again be worthwhile taking the central member's tetrahedral vertex figure and examining the effect of missing out different pairs of opposing polygons.

Solids	E/V	F/V	S/V
truncated trihexatile	4	2 ⬡	2
		2 □	
		(1) ○	
		(1) ‖	
Truncated Trihexatile Triskew			
rhombitrihexatile	5	4 □	1
hexaprism		1 △	1
cube		(1) ⬡	2
triprism		(2) □	1
Rhombitrihexaprism Triskew			
rhombitrihexatile	6	4 □	2
		2 △	
		(1) ⬡	
		(2) ‖	
Rhombitrihexatile Triskew			

Table 21: Further Degenerate Triskews

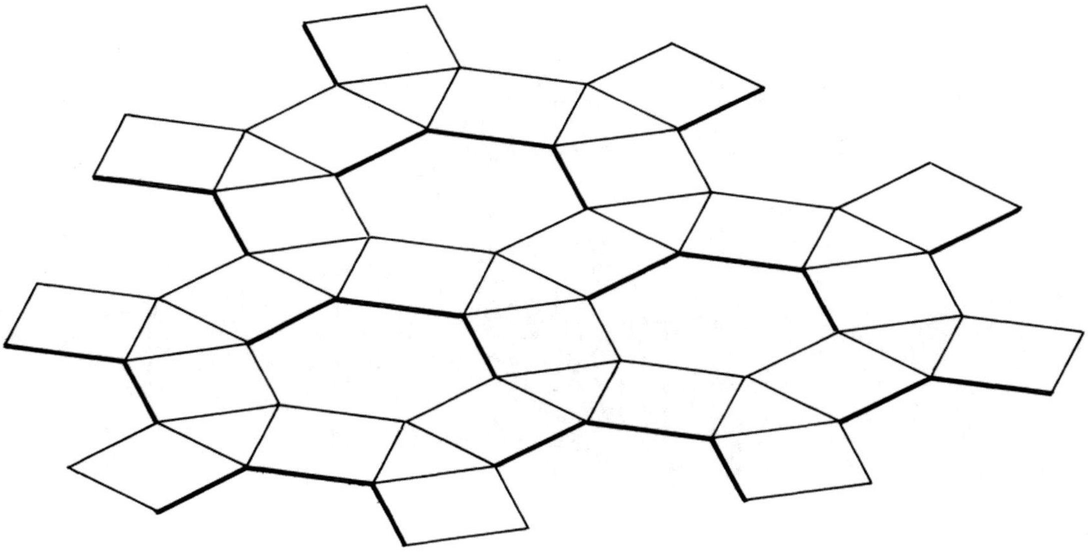

Rhombitrihexatile Triskew $[t^2t^2\{6,3,2\}]$

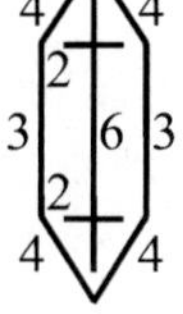

Fig.51: Another Resultant Degenerate

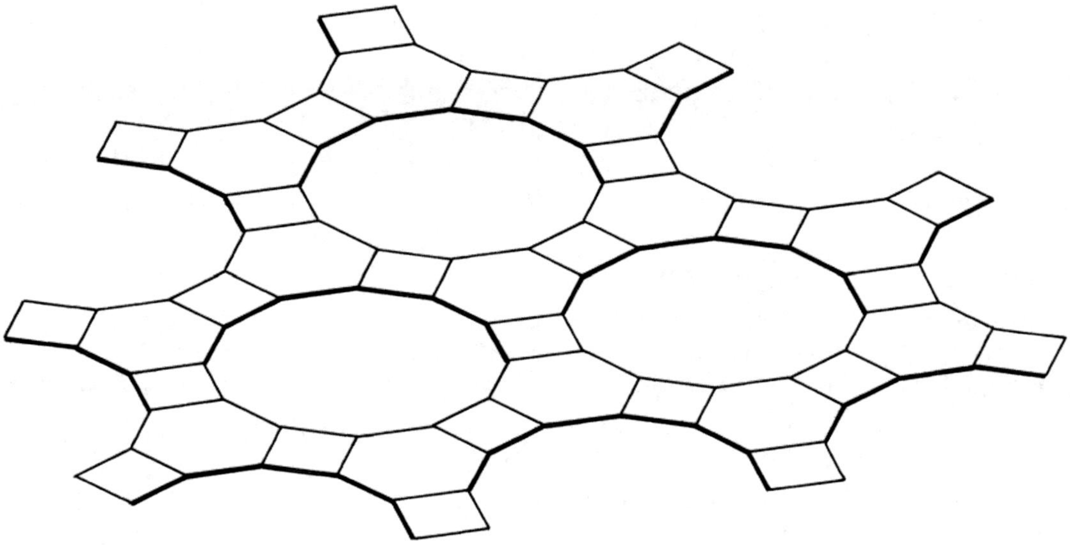

$[tt^2\{6,3,2\}]$

Truncated Trihexatile Triskew

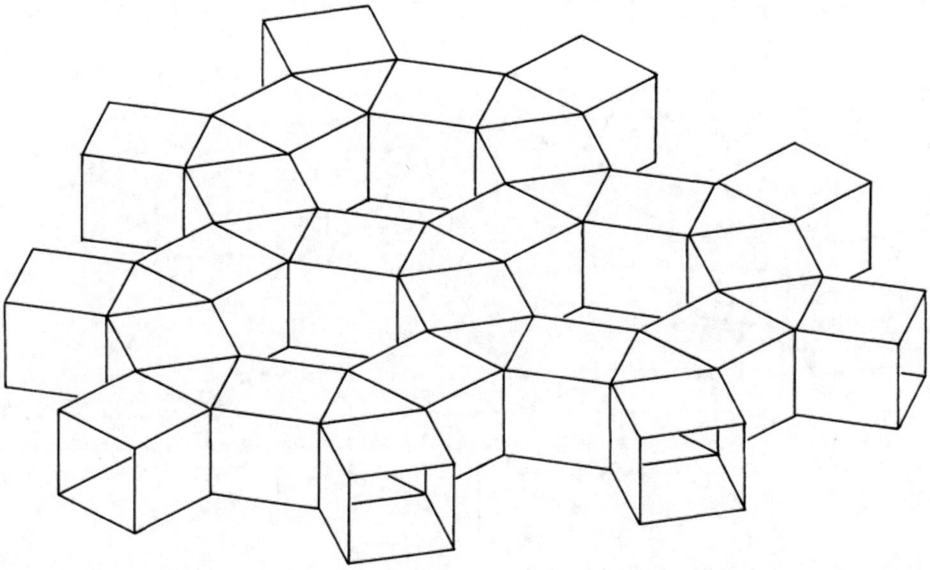

$[o^2\{6,3,2\}]$

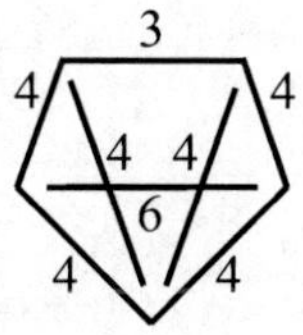

Rhombitrihexaprism Triskew

Fig.52: More Intermediate Triskews

FURTHER DEGENERATES

One way, missing out a hexagon and a square, leads to the central degenerate skew polyhedron shown below in Figure 53. This has a vertex with a dodecagon, a square and two squares present.

Carrying out vertex transformations on this we can arrive at a further pair of 'parent' skew polyhedra, as shown opposite in Figure 54. One of these comprises back to back dodecagons with triangular holes and can be assigned the symbol {12,4 |3} in Coxeter's notation. The other is conveniently its dual {4,12 |3} comprising twelve back to back squares at each vertex in a triangular lattice pattern.

All this family will prove to be 6-connected around this {3,6} lattice pattern.

Solids	E/V	F/V	S/V
tritile	12	12 □	1
triprism		(6) △	6
{4,12 |3}		(6) ‖	
Triprism Hexaskew			
truncated hexatile	4	4 ○	2
		(1) △	
{12,4 |3}		(1) ‖	
Truncated Hexatile Hexaskew			
truncated trihexatile	4	1 ○	1
dodecaprism		3 □	1
hexaprism		(1) ⬡	1
cube		(1) □	1
Truncated Trihexaprism Hexaskew			

Table 22: Degenerate Hexaskews

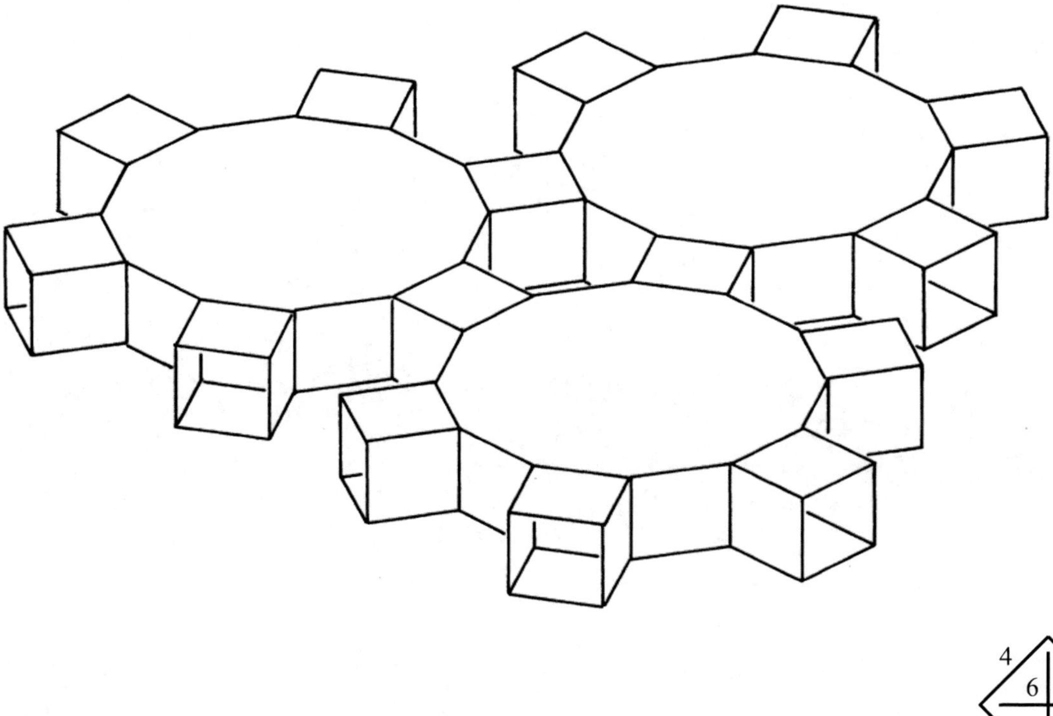

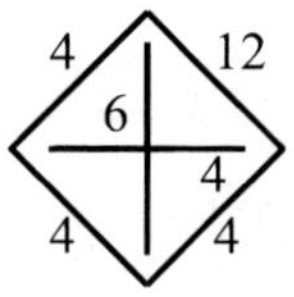

Truncated Trihexaprism Hexaskew [o{3,6,2}]

Fig.53: Another Central Degenerate

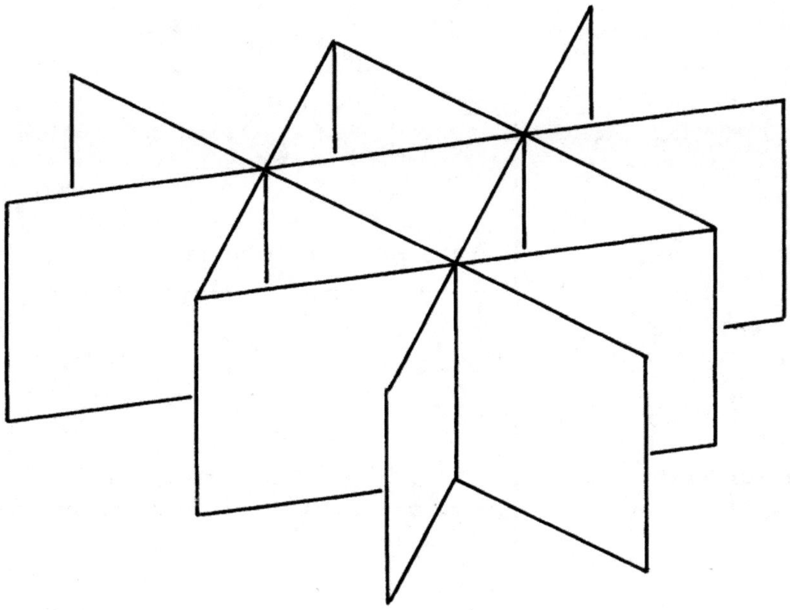

{4,12 |3}
Triprism Hexaskew

[o³{3,6,2}]

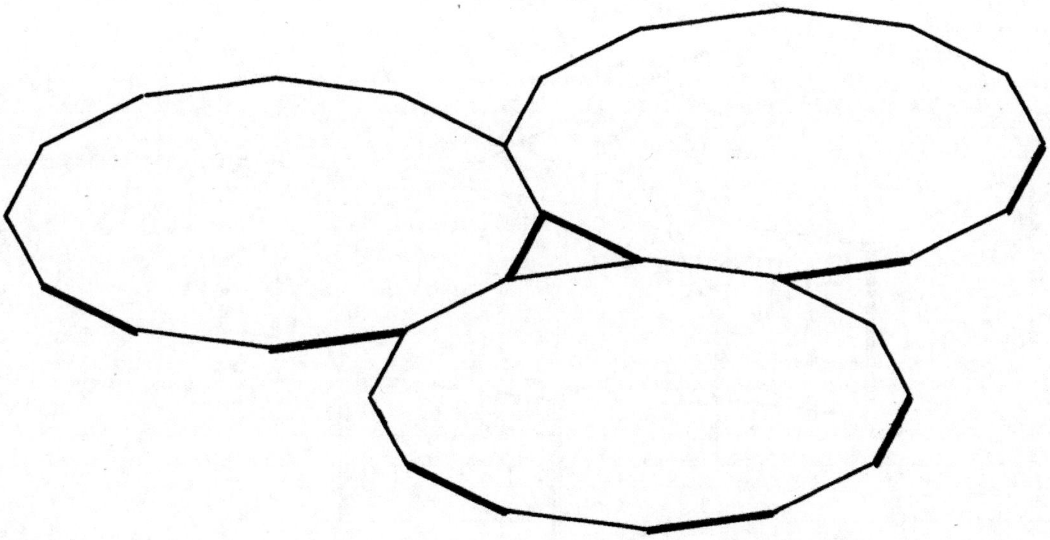

{12,4 |3}
Truncated Hexatile Hexaskew

[t³{3,6,2}]

Fig.54: Parent Degenerate Hexaskews

MORE DEGENERATE HEXASKEWS

Like the {4,6 |6} family, the central member of this {4,12 |3} family's grid has an asymmetrical vertex figure with three squares and a dodecagon around a missing square and hexagon.

Again we find that there are two sets of intermediates and resultant, each employing the same polygons at each vertex, but arranged in two different ways, depending in which direction the vertex is folded back.

Opposite and below in Figures 56 and 55 are one pair of these intermediates and one resultant.

Solids	E/V	F/V	S/V
truncated hexatile	4	2 ○	1
dodecaprism		2 □	2
triprism		(1) □	1
		(1) △	
Truncated Hexaprism Hexaskew			
hexatile	5	1 ⬡	1
hexaprism		4 □	3
		(2) ⬡	
		(1) ∥	
Truncated Triprism Hexaskew			
trihexatile	6	2 ⬡	1
hexaprism		4 □	2
triprism		(2) △	2
		(1) ∥	
Trihexaprism Hexaskew			

Table 23: More Degenerate Hexaskews

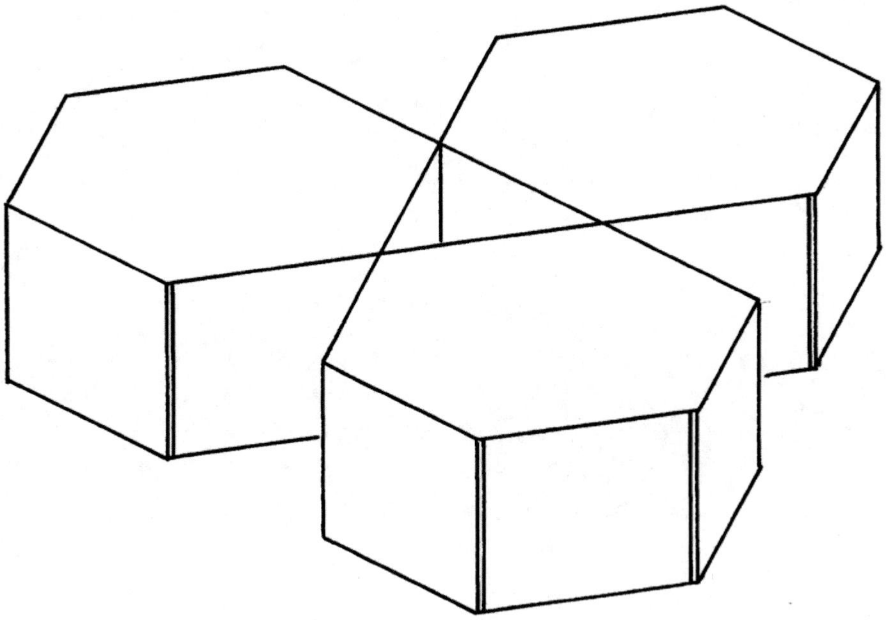

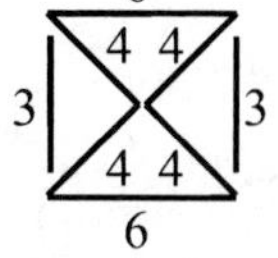

Trihexaprism Hexaskew [$t^2t^2\{2,6,3\}$]

Fig.55: Resultant Degenerate Hexaskew

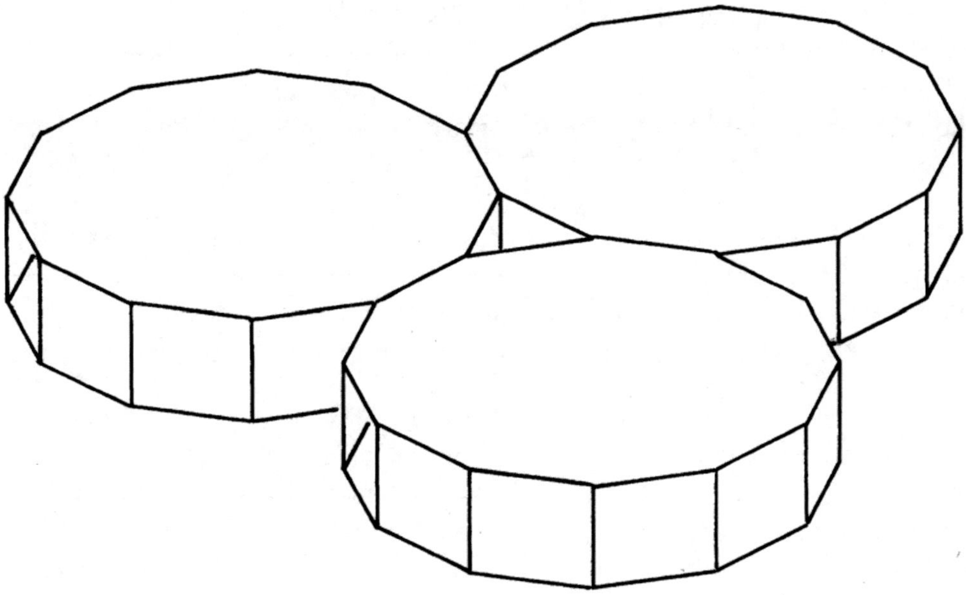

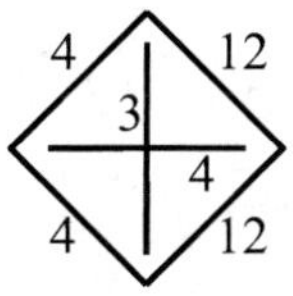

Truncated Hexaprism Hexaskew $[tt^2\{2,6,3\}]$

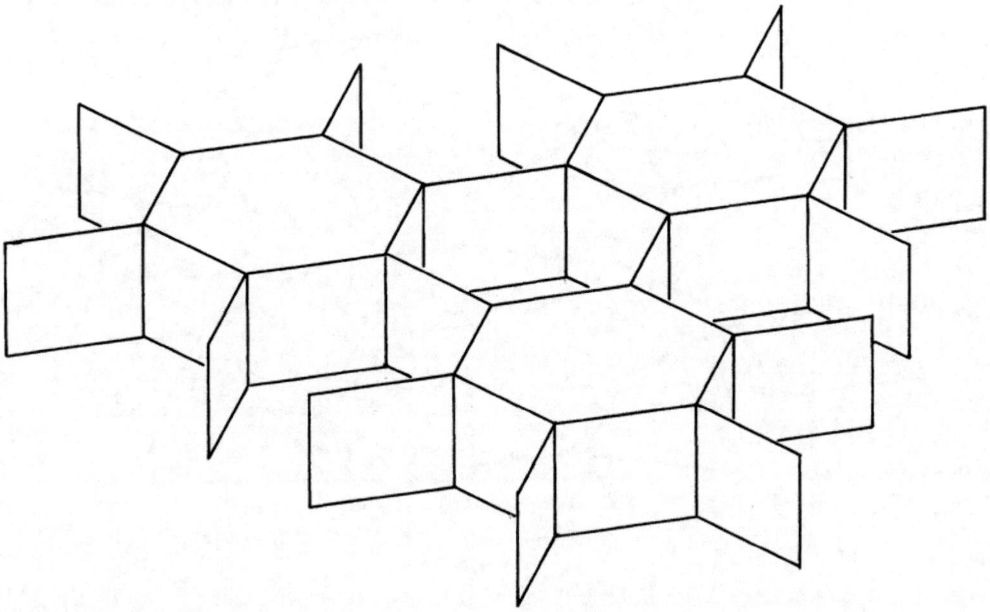

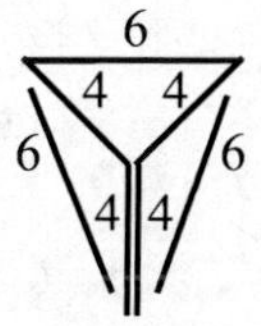

Truncated Triprism Hexaskew $[o^2\{2,6,3\}]$

Fig.56: Intermediate Degenerate Hexaskews

FURTHER DEGENERATES

Following on as before to find the other members of this family grid there are a second pair of intermediates and another resultant as shown opposite and below in Figures 58 and 57.

Comparing these one to one with the first set of such in Figures 56 and 55, they can again be seen to exhibit exactly the same polygons at each vertex, but arranged differently.

With this second grid now filled in, it will again be worthwhile taking the central member's tetrahedral vertex figure and turning it to its third orientation, this time missing out two opposing squares.

First, however, we will quickly recap where we have got to with the degenerate skew polyhedra.

Solids	E/V	F/V	S/V
truncated trihexatile	4	2 ⬡	2
		2 □	
		(1) ⬡	
		(1) ∥	
Truncated Trihexatile Hexaskew			
rhombitrihexatile	5	1 ⬡	1
hexaprism		4 □	1
cube		(2) □	2
triprism		(1) △	1
Rhombitrihexaprism Hexaskew			
rhombitrihexatile	6	2 ⬡	2
		4 □	
		(1) △	
		(2) ∥	
Rhombitrihexatile Hexaskew			

Table 24: Further Degenerate Hexaskews

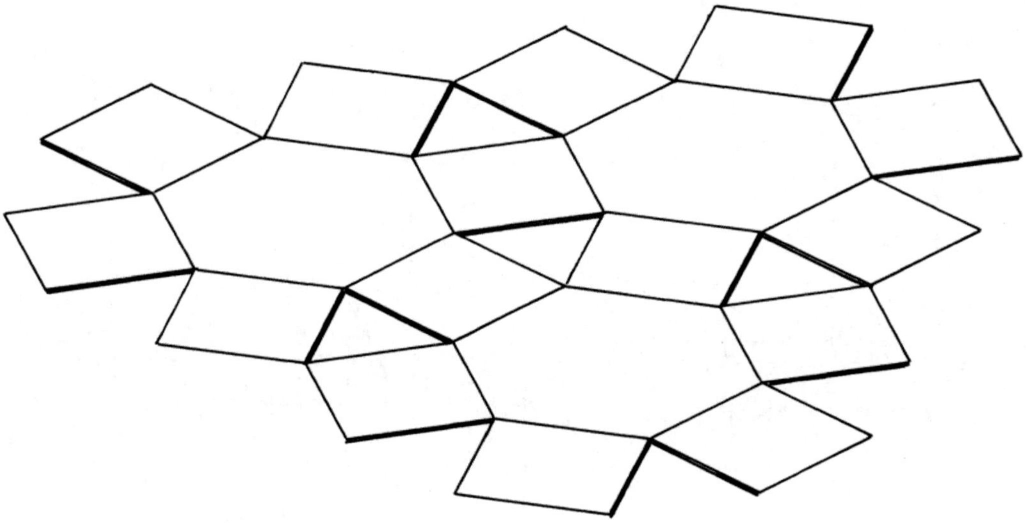

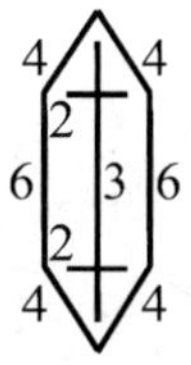

Rhombitrihexatile Hexaskew $[t^2t^2\{3,6,2\}]$

Fig.57: Another Resultant Degenerate

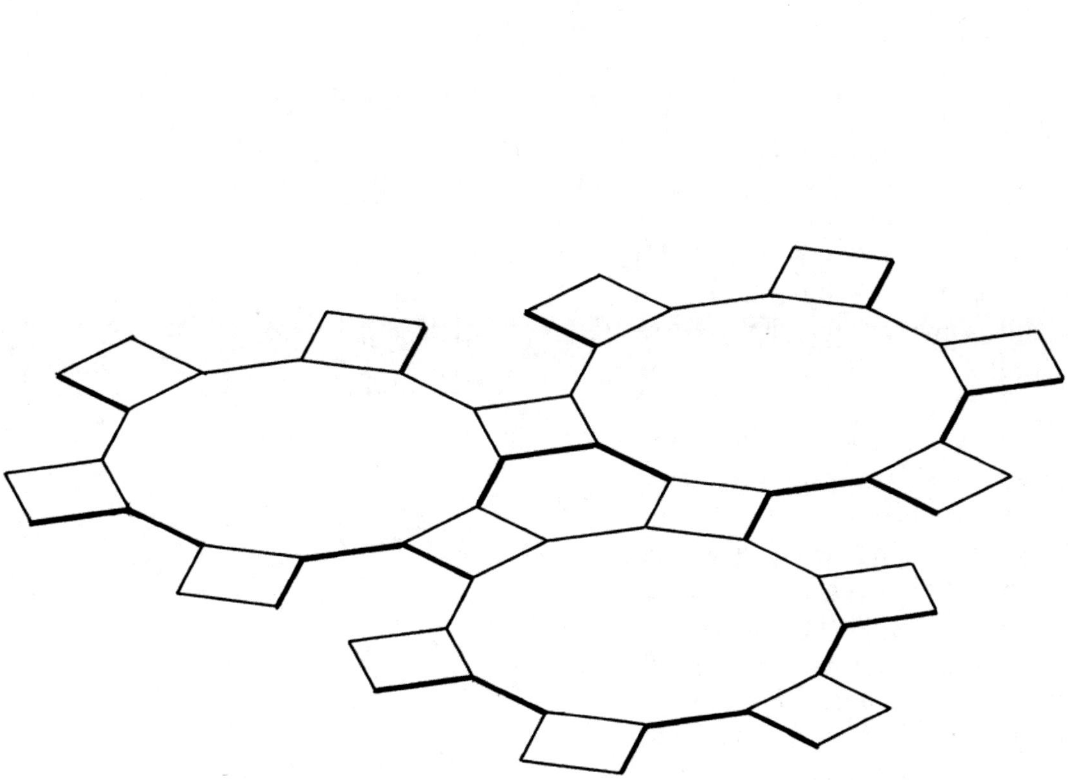

Truncated Trihexatile Hexaskew $[tt^2\{3,6,2\}]$

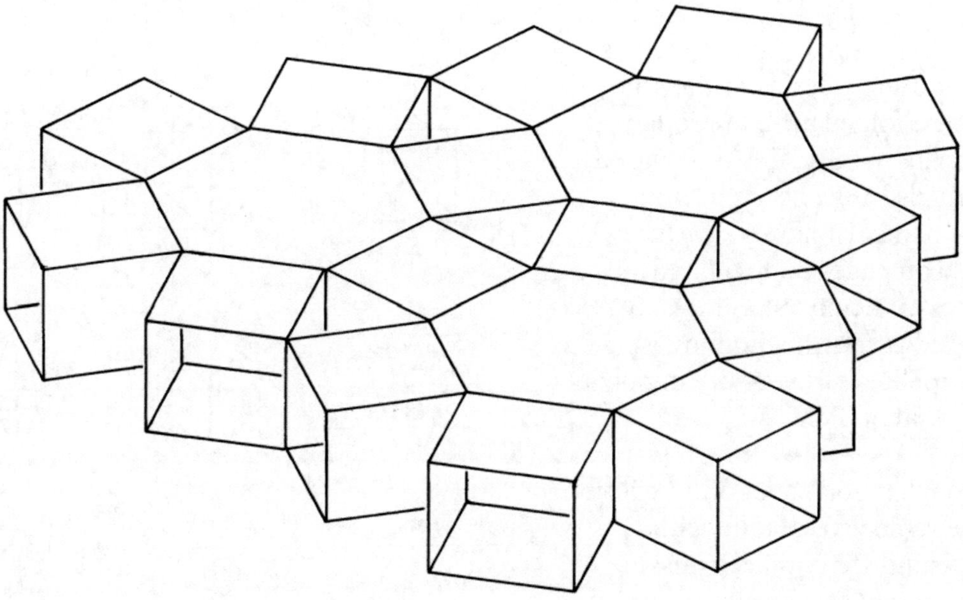

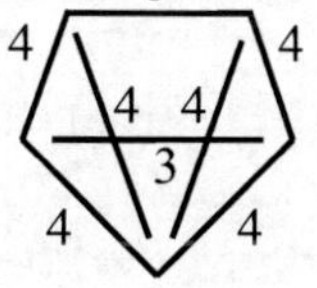

Rhombitrihexaprism Hexaskew $[o^2\{3,6,2\}]$

Fig.58: More Intermediate Hexaskews

DEGENERATE FAMILIES

The first degenerate family we looked at back on pages 46 to 51 was produced by looking for a {4,8} family, related to the {8,8} in the same way that the {4,6} relates to the {6,6}.

What followed was a 4-connected family that stayed in two dimensions, rather than folding up into three. However it was not totally two dimensional, some members having a minimal thickness in the third dimension. The 4-connected spaces contained by these can be seen to follow truncations of the square tiling {4,4}, when it is considered as a three dimensional polytope {4,4,2} that includes prisms amongst its building blocks. These correspondences have been noted in the figures for this family [ ], the grid for which is shown adjoining in Figure 59, whilst Plate 8 opposite shows photographs of models of the members of this family.

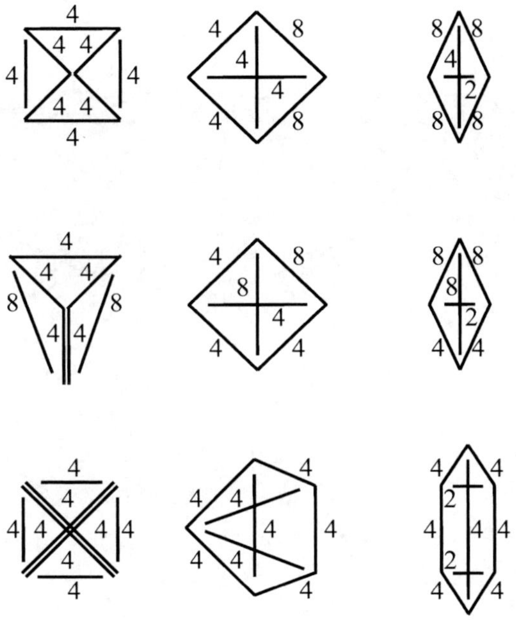

Fig.59: {4,8} Degenerate Tetraskew Grid

Success with describing this family led us on to consider different forms of the {4,6} family's vertices, folded back to produce another degenerate family, this time based around truncations of the hexagon tiling {6,3}. This 3-connected family of triskews was considered on pages 52 to 57 and similarly relates to members of the {6,3,2} polytope. Its grid is shown adjoining in Figure 60 and models of its family members are shown opposite in Plate 9.

Moving on from here on pages 58 to 63 we considered the same central tetrahedral vertex figure with different faces missing and found the third of the regular tilings, the triangular {3,6}, in the form of the polytope {3,6,2}. This provided the basis for a {4,12} family of degenerate skew polyhedra, the grid for which is shown in Figure 61, facing its family members in Plate 10.

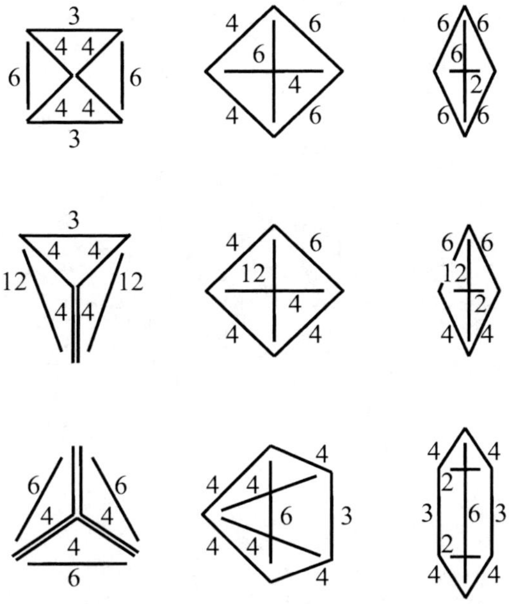

Fig.60: {4,6} Degenerate Triskew Grid

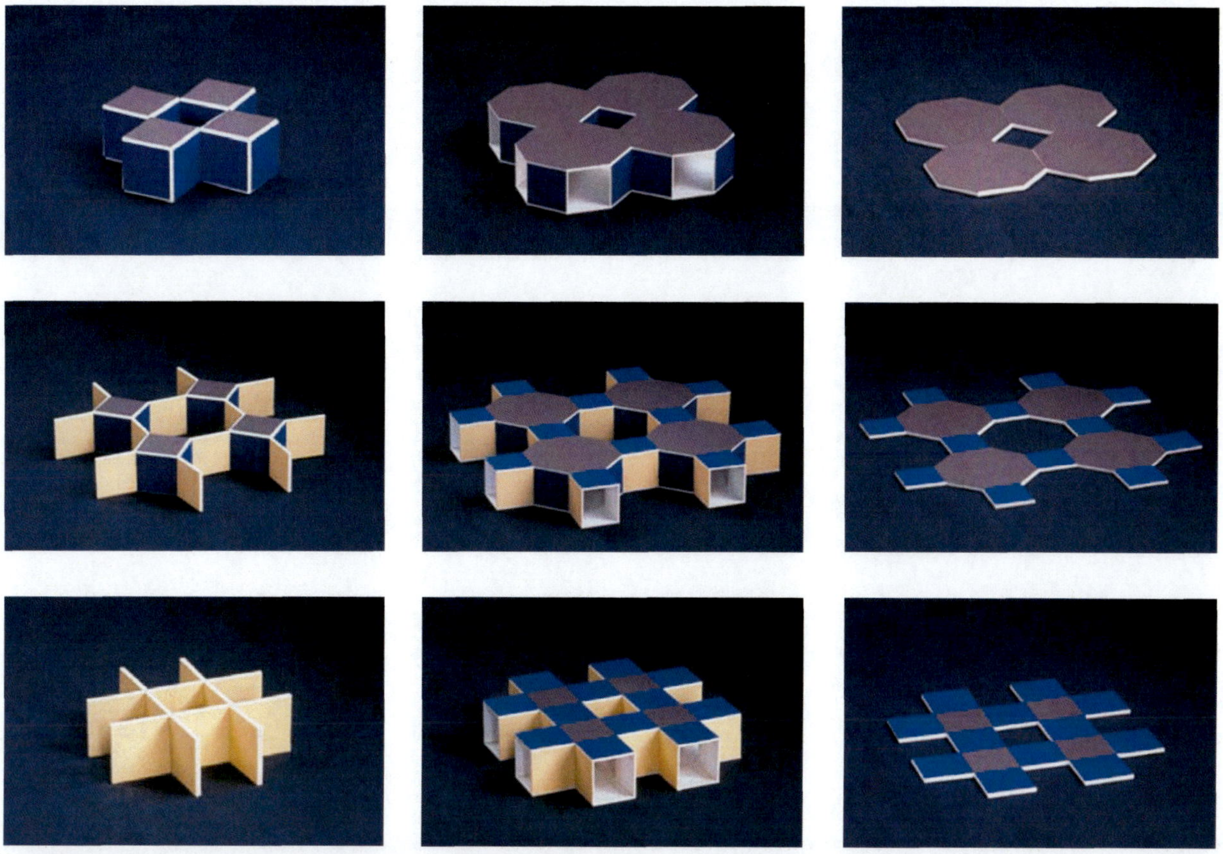

Plate 8: {4,8} Degenerate Tetraskew Family [2-4,2-∞]

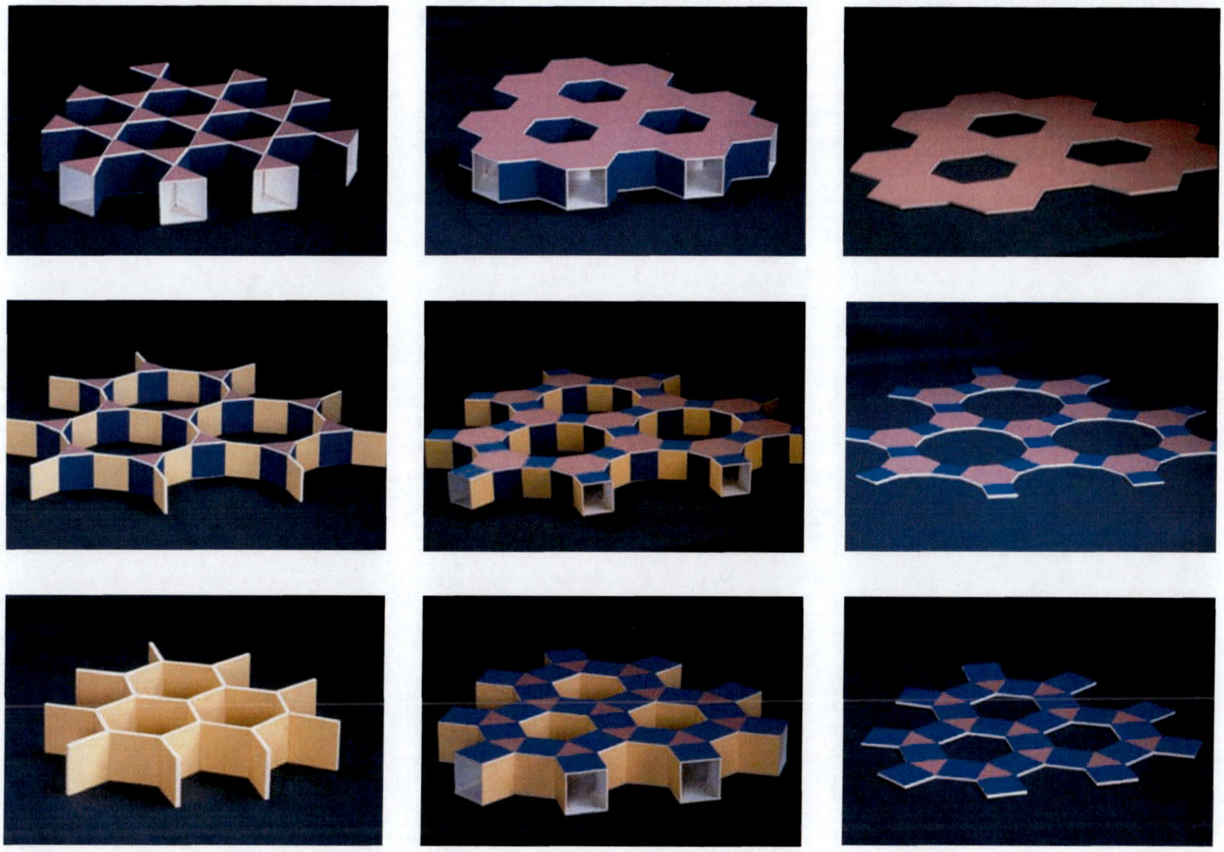

Plate 9: {4,6} Degenerate Triskew Family [2-3,2-∞]

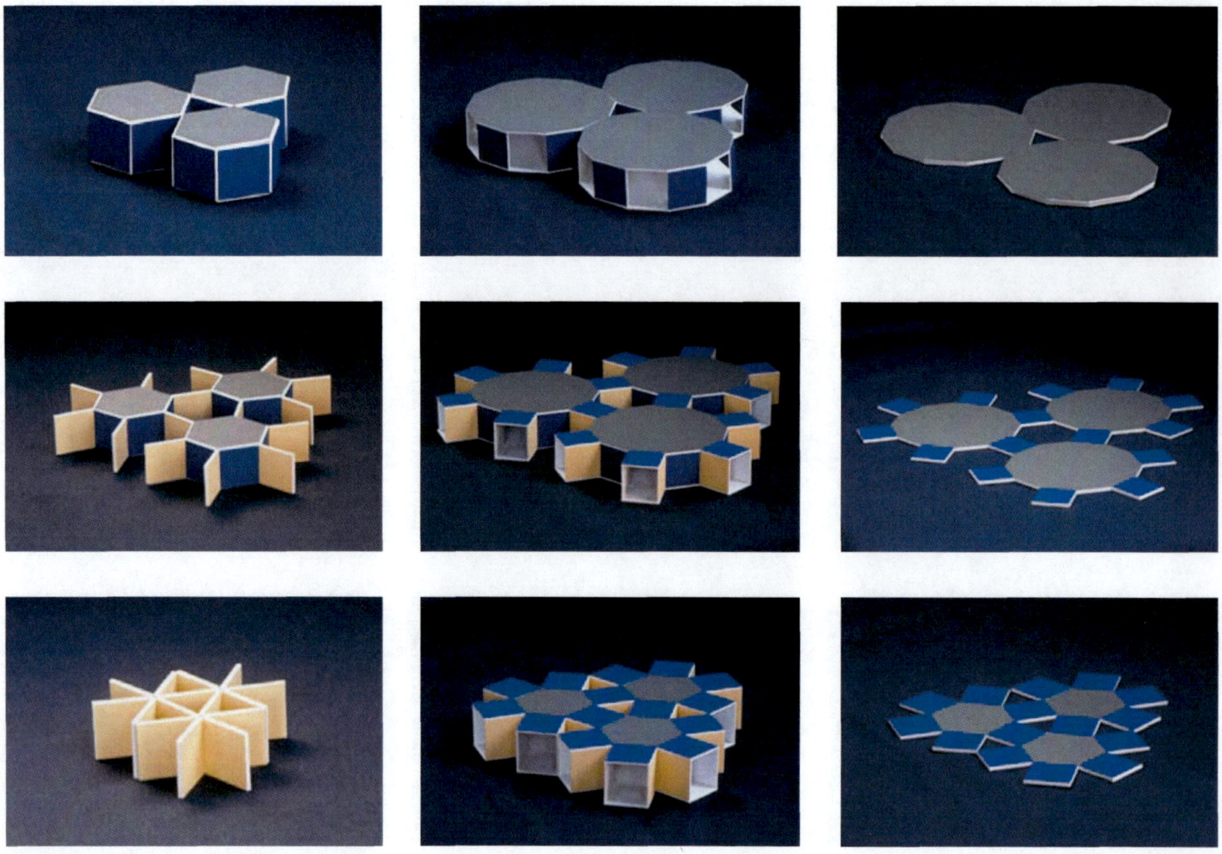

Plate 10: {4,12} Degenerate Hexaskew Family [2-6,2-∞]

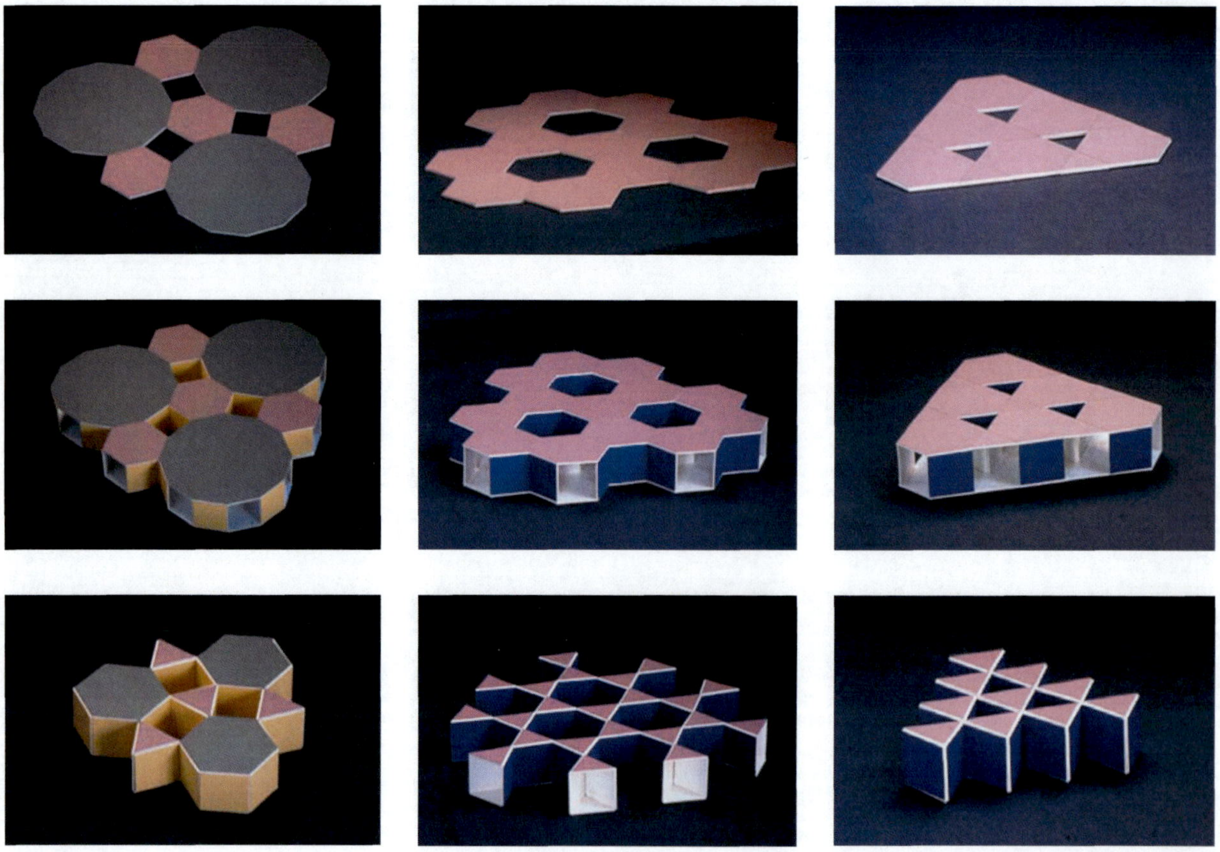

Plate 11: Degenerate Families [3-6,2-∞] and [3-3,2-∞]]

ANOTHER DEGENERATE FAMILY

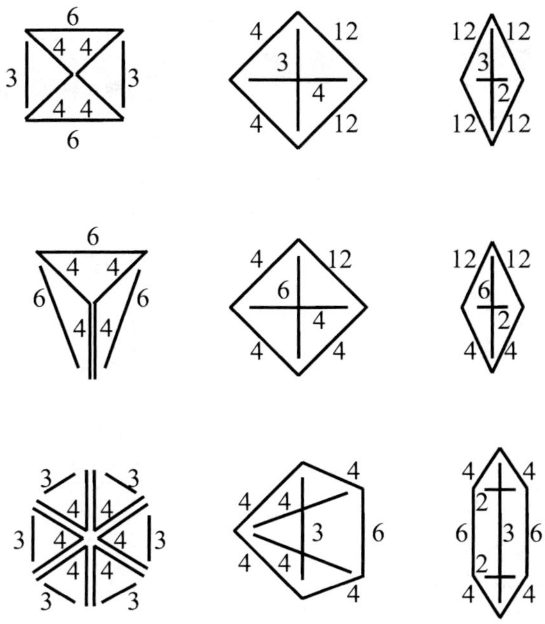

Fig.61: {4,12} Degenerate Hexaskew Grid

Taking the {4,6} or {4,12} central member's tetrahedral vertex figure, we can miss out the other pair of polygons, two squares, to get a third central member, which will start off another family grid.

This is shown below in Figure 62 and can be seen to generate a whole family, all 3- and 6-connected. Two of these are tilings with slots and four are packings of prisms with slots. Only the three in the central column actually have polygonal holes worthy of a model and are shown opposite in the left hand column of Plate 11.

The remainder of Plate 11 is taken up with the results of taking one of the {4,6} intermediates with a tetrahedral vertex figure and treating it as the central member of a grid in its own right shown below in Figure 63. The missing left hand column in the plate simply duplicates the one on the right.

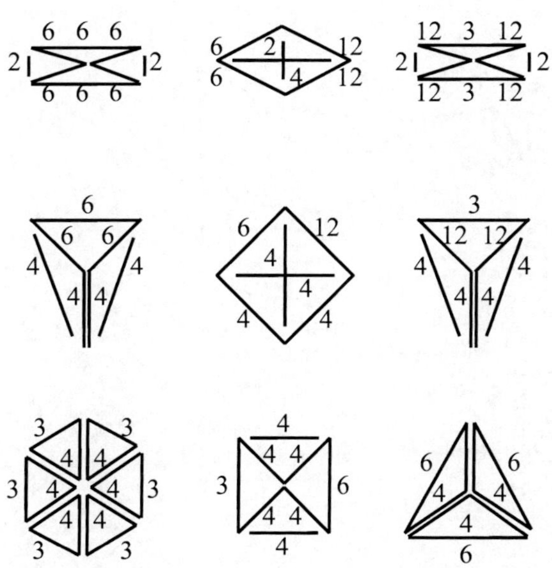

Fig.62: Degenerate [3-6] Family Grid

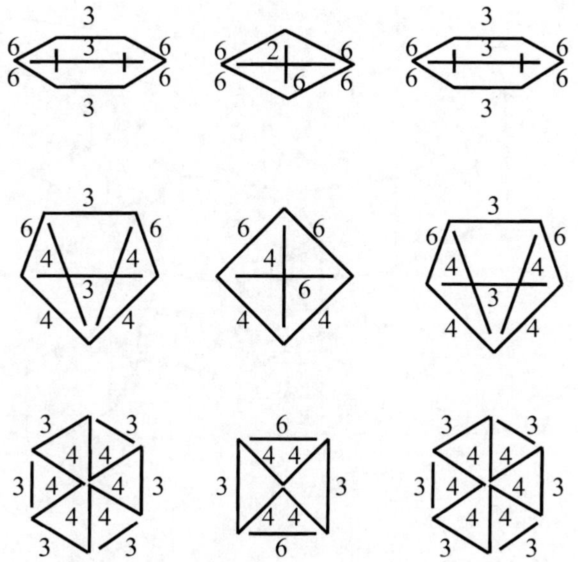

Fig.63: Degenerate [3-3] Family Grid

DEGENERATE TRIHEXASKEWS

Going back to our {4,6} and {4,12} central member with their tetrahedral vertex figures, the third way of aligning these, missing out two squares, leads to the central degenerate skew polyhedron shown below in Figure 64. This has a vertex with a dodecagon, a hexagon and two squares present.

Carrying out our tried and tested vertex transformations on this we can arrive at further skew polyhedra. The first two intermediates of these, which appeared top and bottom left in Plate 11, are shown opposite in Figure 65.

All this family will prove to be both 3- and 6-connected around the lattice pattern of a triangle-hexagon tiling (trihexatile).

Solids	E/V	F/V	S/V
truncated trihexatile	4	2 ◯	2
		2 ⬡	
		(1) □	
		(1) ‖	
Truncated Trihexatile Trihexaskew			
rhombitrihexatile	6	1 ⬡	1
hexaprism		4 □	1
cube		1 △	2
triprism		(2) □	1
		(1) ‖	
Rhombitrihexaprism Trihexaskew			
truncated trihexatile	4	1 ◯	1
dodecaprism		1 ⬡	1
hexaprism		2 □	1
cube		(2) □	1
Truncated Trihexaprism Trihexaskew			

Table 25: Degenerate Trihexaskews

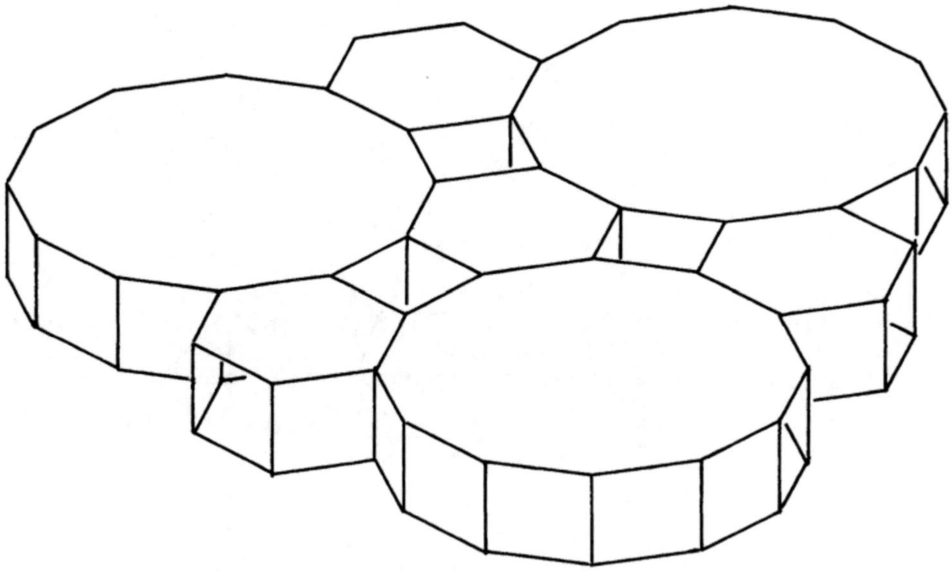

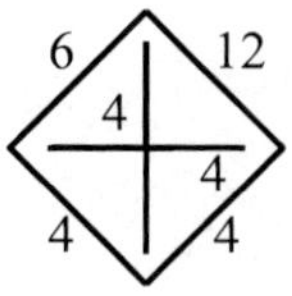

Truncated Trihexaprism Trihexaskew [o{6,3,2}] [o{3,6,2}]

Fig.64: Another Central Degenerate

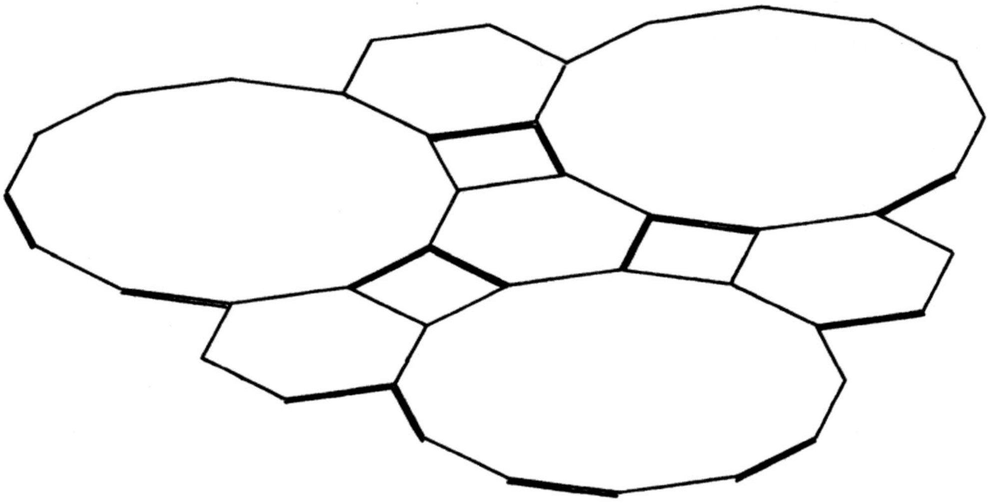

Truncated Trihexatile Trihexaskew $[tt^2\{6,3,2\}]$ $[tt^2\{3,6,2\}]$

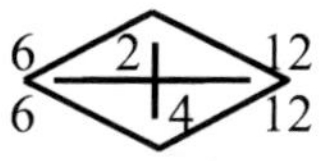

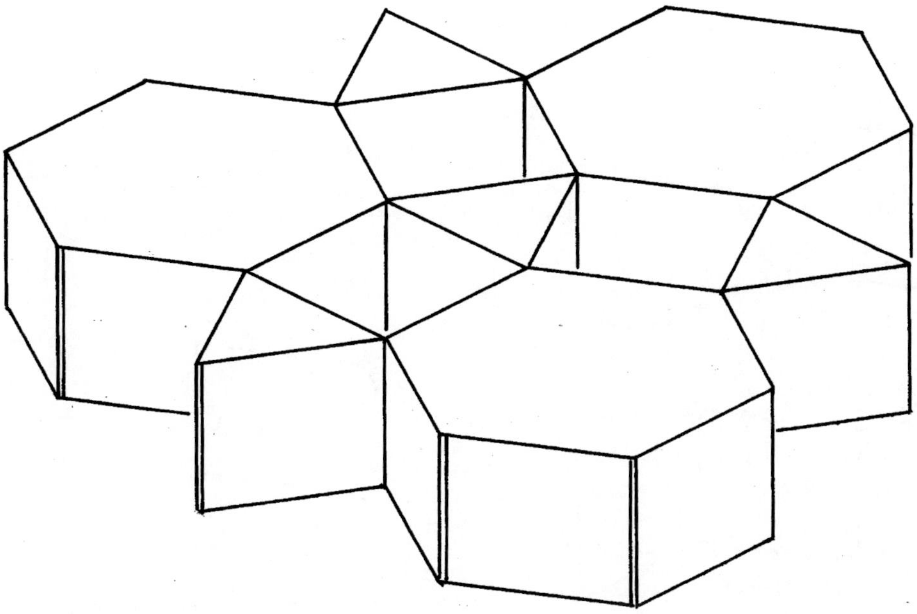

Rhombitrihexaprism Trihexaskew $[o^2\{6,3,2\}]$ $[o^2\{3,6,2\}]$

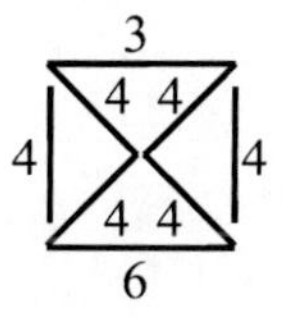

Fig.65: More Intermediates

MORE TRIHEXASKEWS

Looking at this new [3-6] family grid we will now describe the members in the left hand column. Centrally in this column there is a further intermediate as shown below in Figure 66.

It is comprised entirely of hexagon prisms packed together, half are reduced dodecagons, the other half are the original hexagon prisms of the central member of this grid. These last prisms have slots in between them, which have originated from the original cubic holes.

Opposite in Figure 67 are the two resultants at the top and bottom of this left hand column. One is a hexagon tiling with slots, the other a collection of triangular prisms with slots.

Solids	E/V	F/V	S/V
hexatile	6	6 ⬡	2
		(3) ‖	
Truncated Tritile Trihexaskew			
tritile	18	18 △	1
triprism		(6) ‖	6
Triprism Trihexaskew			
hexatile	5	3 ⬡	1
hexaprism		2 □	3
		(2) □	
		(1) ‖	
Truncated Triprism Trihexaskew			

Table 26: Degenerate Trihexaskews

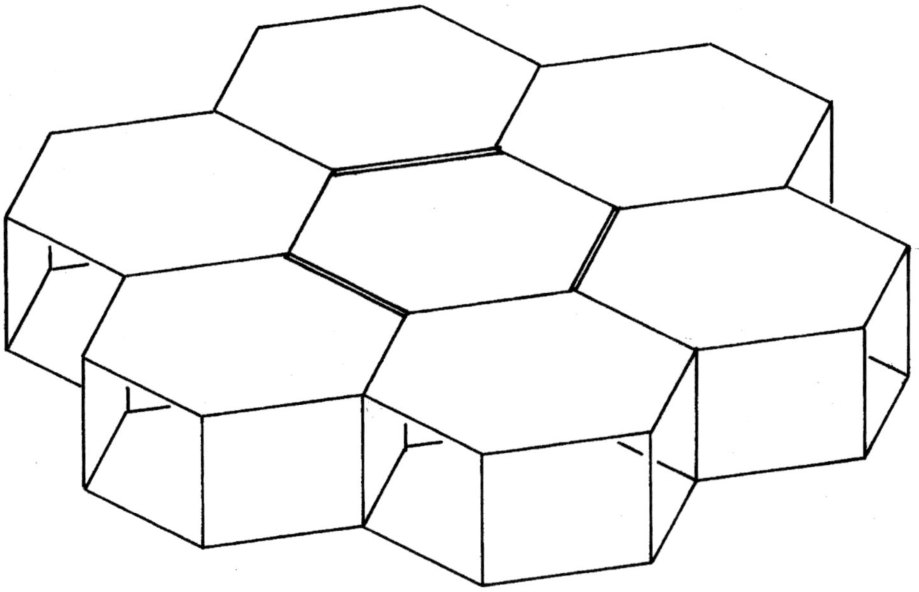

Truncated Triprism Trihexaskew $[tt^2\{2,3,6\}]\ [o^2\{2,6,3\}]$

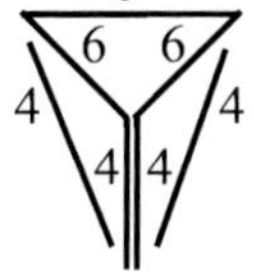

Fig.66: Another Intermediate Trihexaskew

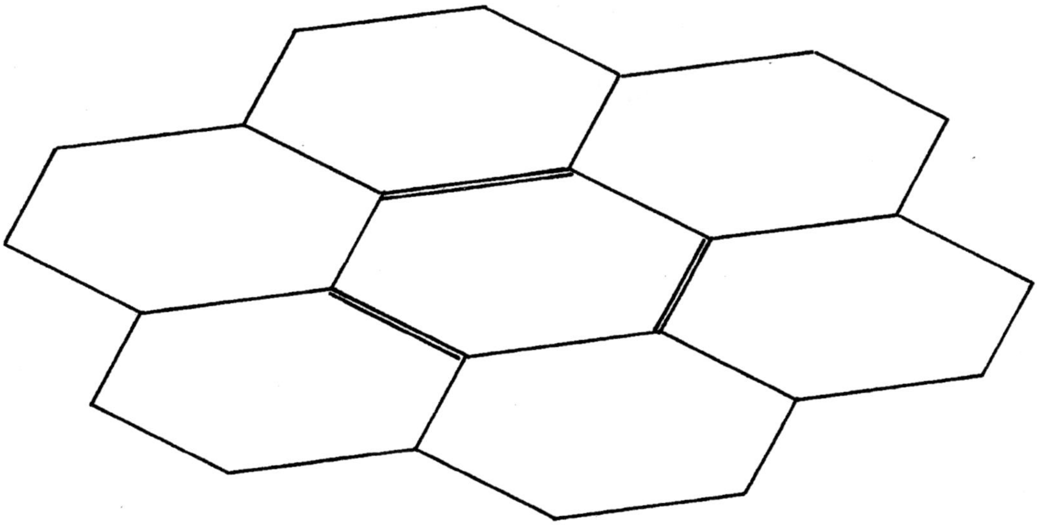

Truncated Tritile Trihexaskew $[t^3\{6,3,2\}]$

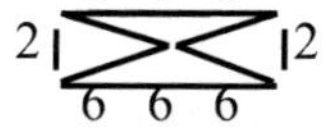

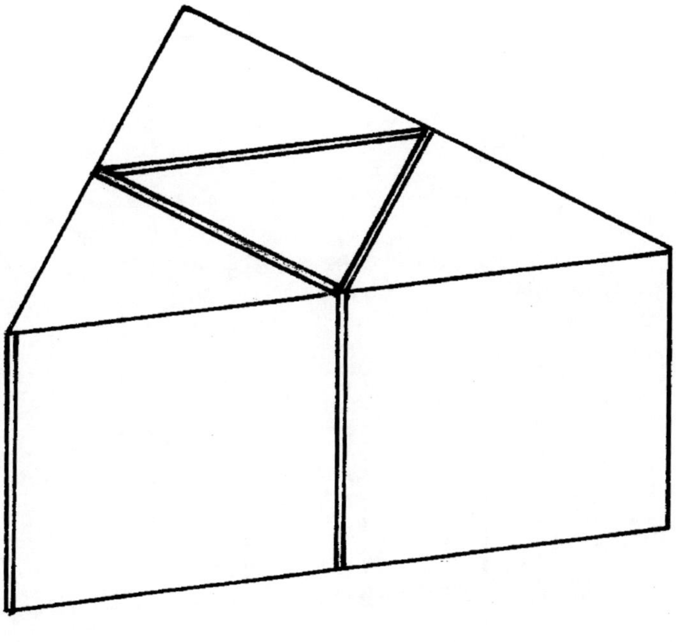

Triprism Trihexaskew $[o^3\{3,6,2\}]$

Fig.67: Resultant Trihexaskews

FURTHER TRIHEXASKEWS

Finally on this family grid we will now describe the members in the right hand column. Centrally in this column there is a further intermediate as shown below in Figure 68.

It is comprised entirely of dodecagon and triangle prisms packed together, the latter derived from the original hexagon prisms of the central member of this grid. The dodecagon prisms have slots in between them, which have originated from the original cubic holes, now squeezed at right angles to the way they were before.

Opposite in Figure 69 are the remaining two resultants at the top and bottom of this right hand column. One is a dodecagon-triangle tiling with slots, the other another collection of hexagon prisms with slots, this time all derived from the original dodecagons.

Solids	E/V	F/V	S/V
truncated hexatile	6	4 ○	2
		2 △	
		(3) ‖	
Truncated Hexatile Trihexaskew			
hexatile	9	3 ⬡	1
hexaprism		6 □	3
		(3) ‖	
Hexaprism Trihexaskew			
truncated hexatile	5	2 ○	1
dodecaprism		2 □	2
triprism		1 △	1
		(2) □	
		(1) ‖	
Truncated Hexaprism Trihexaskew			

Table 27: Degenerate Trihexaskews

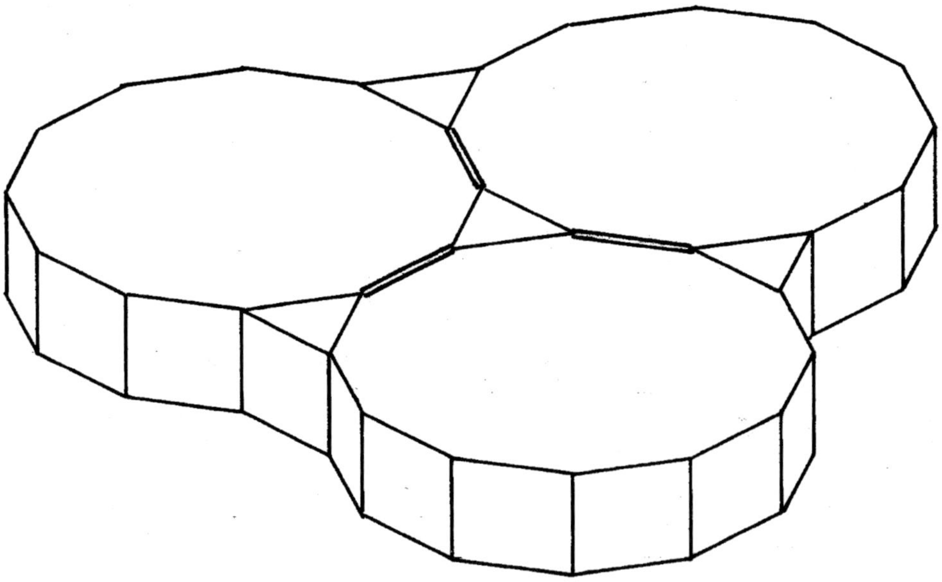

Truncated Hexaprism Trihexaskew [o^2{2,3,6}] [tt^2{2,6,3}]

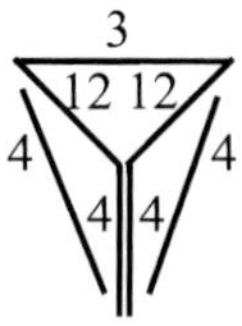

Fig.68: Another Intermediate Trihexaskew

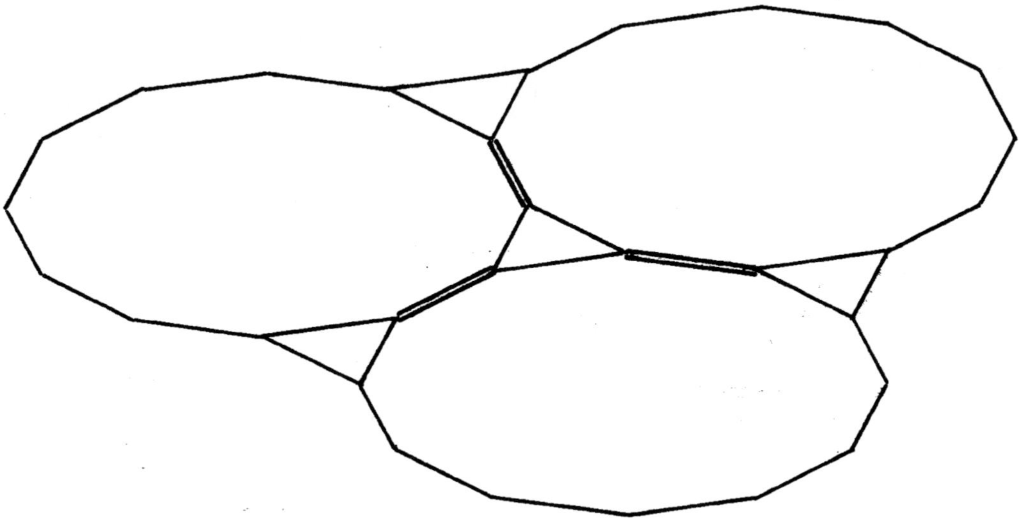

Truncated Hexatile Trihexaskew [t³{3,6,2}]

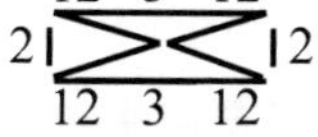

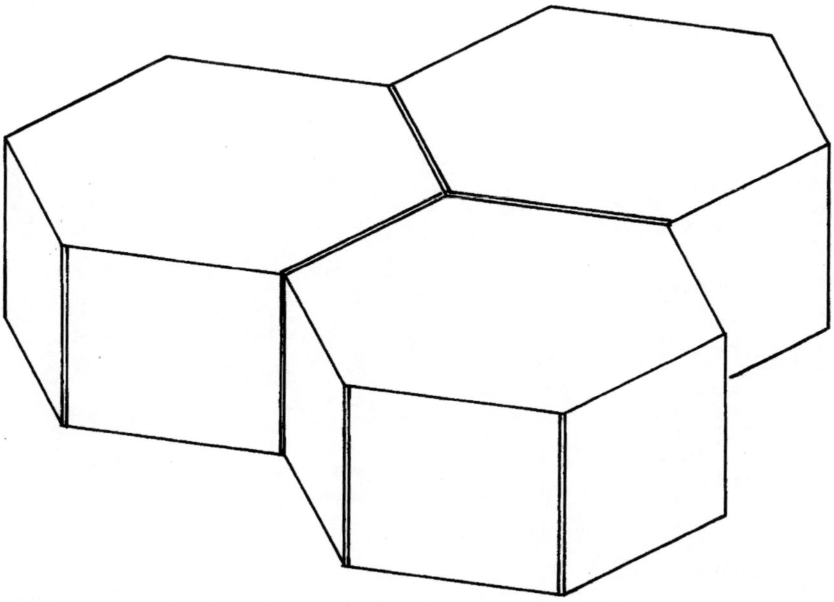

Hexaprism Trihexaskew [o³{6,3,2}]

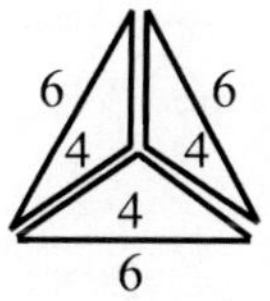

Fig.69: More Resultant Trihexaskews

MORE DEGENERATES

Back with our original skew polyhedra, we saw on page 32 how a further grid for the [6-6,2-6] family was derived from that of the [2-6,2-6] connected {4,6} family by taking as a central member in its own right an intermediate tetrahedral vertex figure. This same operation carried out on the parallel degenerate [2-3] connected {4,6} family produces another grid, that of the totally 3-connected [3-3] family, all of which are folded back versions of their 3-dimensional counterparts.

The central column of this new grid (and Plate 11) is simply the top row of our original {4,6} degenerate grid, as shown in Plate 9. The two adjoining columns are identical and only shown on the right hand side of Plate 11. There are thus only three further degenerate skew polyhedra to be described. The intermediate from the middle of the column is shown below in Figure 70 and the two resultants from top and bottom right or left appear opposite in Figure 71.

Solids	E/V	F/V	S/V
trihexatile	6	4 ⬡	2
		2 △	
		(1) △	
		(2) ∥	
Trihexatile Tritriskew			
tritile	9	6 □	1
triprism		3 △	6
		(3) △	
		(3) ∥	
Triprism Tritriskew			
trihexatile	5	2 ⬡	1
hexaprism		2 □	2
triprism		1 △	2
		(2) □	
		(1) △	
Trihexaprism Tritriskew			

Table 28: Degenerate Tritriskews

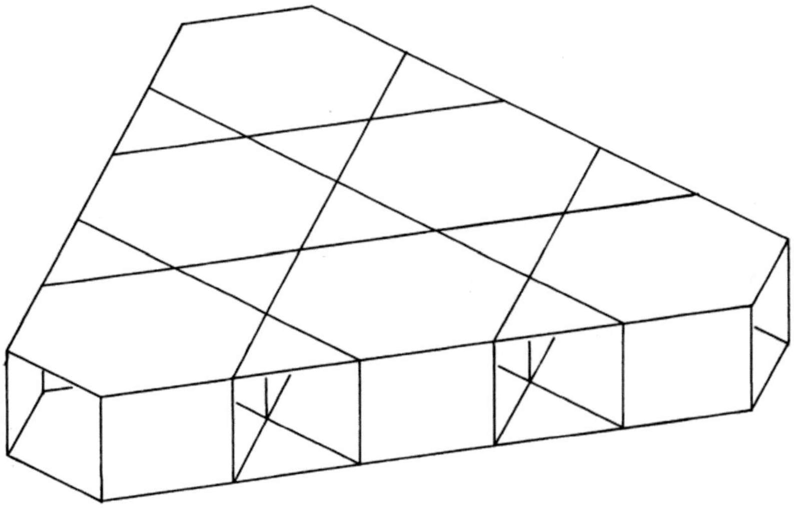

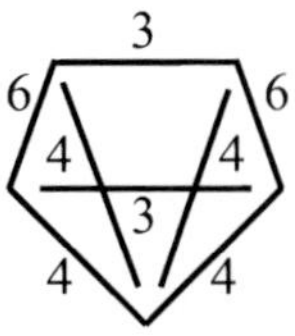

Trihexaprism Tritriskew $[t^2t^2\{2,3,6\}]$ $[t^2t^2\{2,6,3\}]$

Fig.70: Another Intermediate

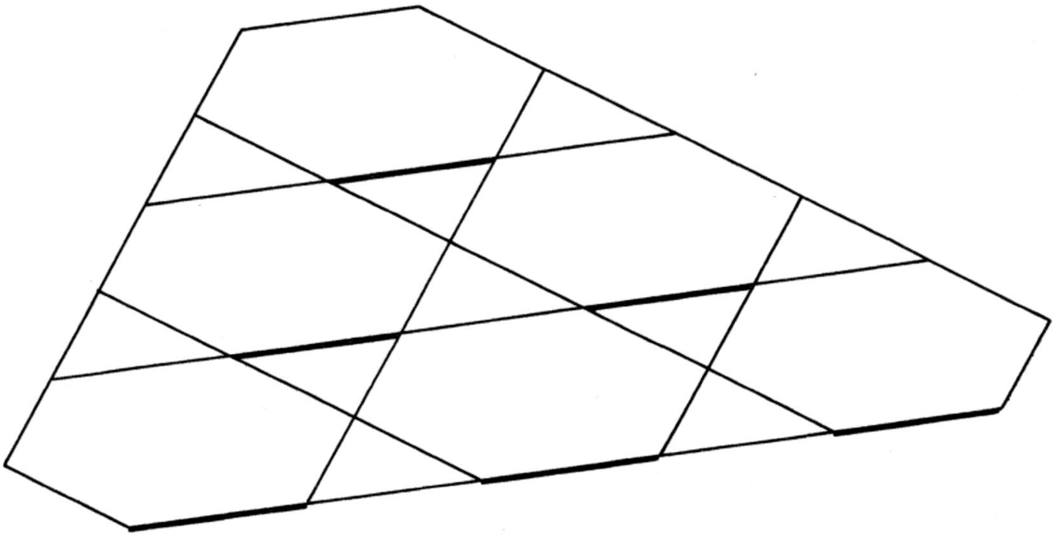

Trihexatile Tritriskew $[t^2\{6,3,2\}]\ [t^2\{3,6,2\}]$

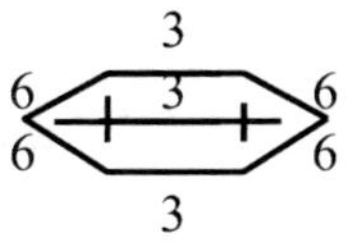

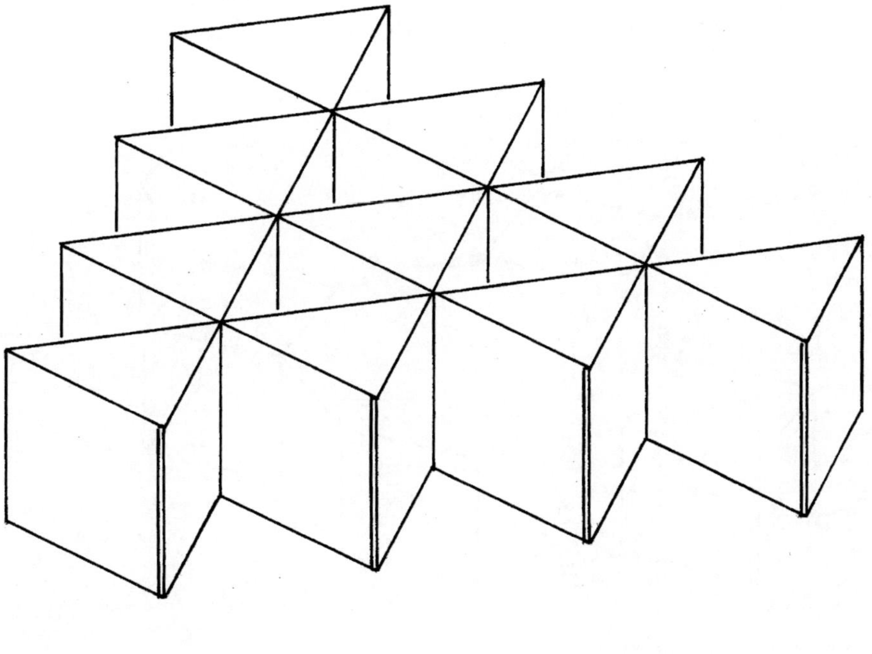

Triprism Tritriskew $[o^3\{3,6,2\}]\ [o^3\{2,6,3\}]$

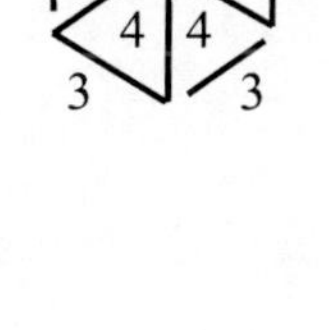

Fig.71: More Resultants

MORE TETRASKEWS

The last few pages have shown that it was worthwhile taking the {4,6} or {4,12} central tetrahedral vertex figures, and looking at them in a different orientation.

With the {4,8} degenerate family, a second way of looking at the central vertex figure repeats what we have already seen because of its inherent symmetry, but a third way produces another grid as shown in Figure 74 following. Again the central column has already been seen as the top row of the original {4,8} degenerate grid and the two adjoining columns are identical.

The three new degenerates are thus an intermediate as shown below in Figure 72 and the two resultants shown opposite in Figure 73. Again these are tilings or assemblies of prisms with slots formed where squares used to be.

Solids	E/V	F/V	S/V
truncated tetratile	6	4 ⯃	2
		2 □	
		(3) ‖	
Truncated Tetratile Tetratetraskew			
tetratile	12	12 □	1
cube		(4) ‖	4
Tetraprism Tetratetraskew			
truncated tetratile	5	2 ⯃	1
octaprism		3 □	2
cube		(2) □	1
		(1) ‖	
Truncated Tetraprism Tetratetraskew			

Table 29: Degenerate Tetratetraskews

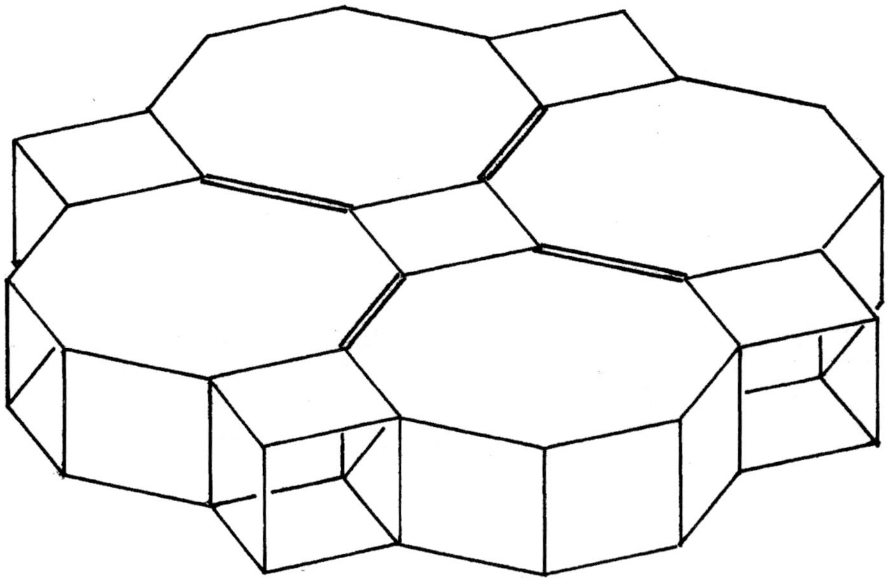

Truncated Tetraprism Tetratetraskew $[o^2\{2,4,4\}]$ $[tt^2\{2,4,4\}]$

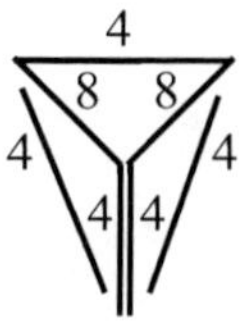

Fig.72: Another Intermediate

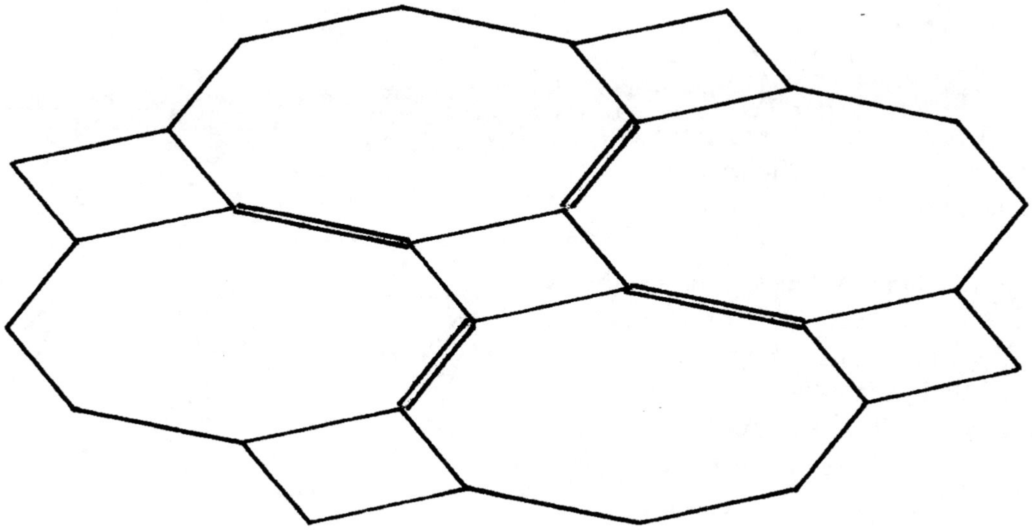

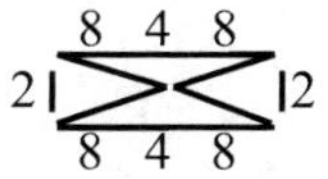

Truncated Tetratile Tetratetraskew [t^3{4,4,2}]

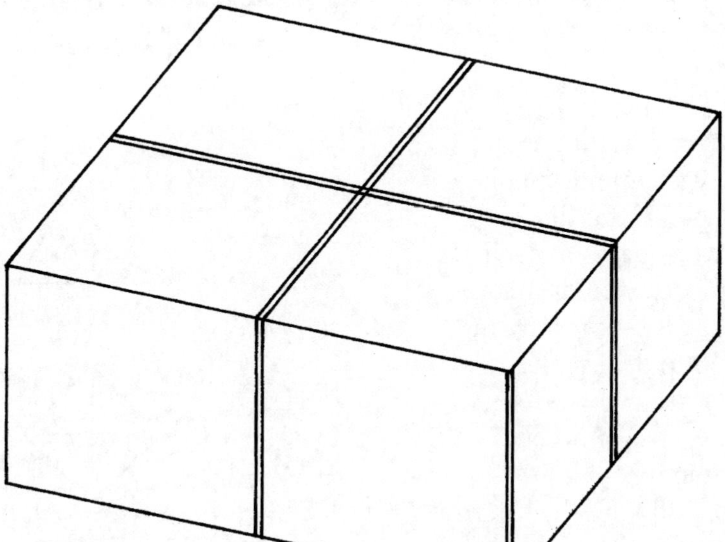

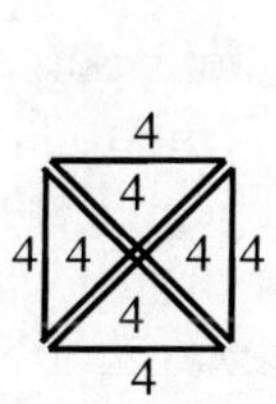

Tetraprism Tetratetraskew [o^3{4,4,2}]

Fig.73: More Resultant Tetraskews

THE LAST TWO FAMILY GRIDS

The family grid covering the degenerate skew polyhedra on the last two pages is shown adjoining in Figure 74. This can be produced either by realigning the [2-4] vertex figure or by taking an intermediate with a tetrahedral vertex figure as the central member of this new grid.

The central column of this grid is simply the top row of our original {4,8} degenerate grid, and all its members are [4-4] connected. In this and the previous [3-3] family it can be seen that both central members and grids are symmetrical about a vertical axis, the central column of each appearing in grids from previously examined families.

In addition it should be noted that these two produce no further tetrahedral vertices that might be taken as starting points for further grids on which we might perform our vertex transformations.

However we have not as yet considered one remaining tetrahedral vertex figure as the centre of a grid. There was an intermediate in the [2-6] connected {4,12} family with two dodecagons and two squares at each vertex, but its missing faces were a square and a triangle, which last is, as we saw in the icosaskews, difficult to deal with.

What we have in Figure 75 is another incomplete grid, this time it is [6-6] connected and derived by taking the top row of the [2-6] {4,12} family as its central row. The triangle faces here prevent the prediction of anything other than the repeated intermediate shown in both top and bottom rows, essentially two back to back hexagon tilings [6-6] connected with holes punched through at the vertices [2-∞].

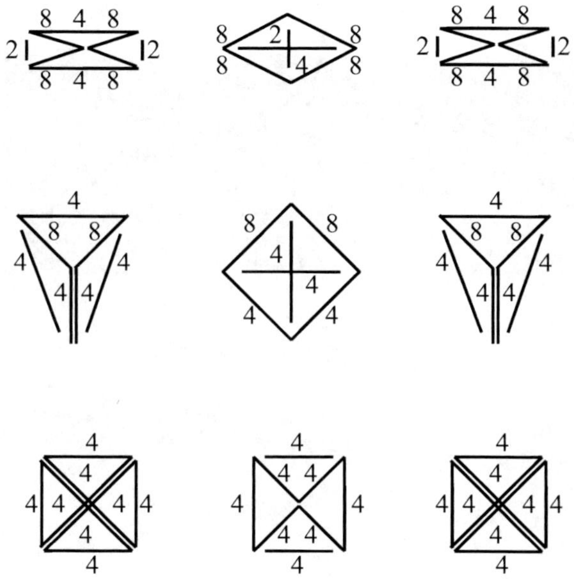

Fig.74: Degenerate [4-4] Family Grid

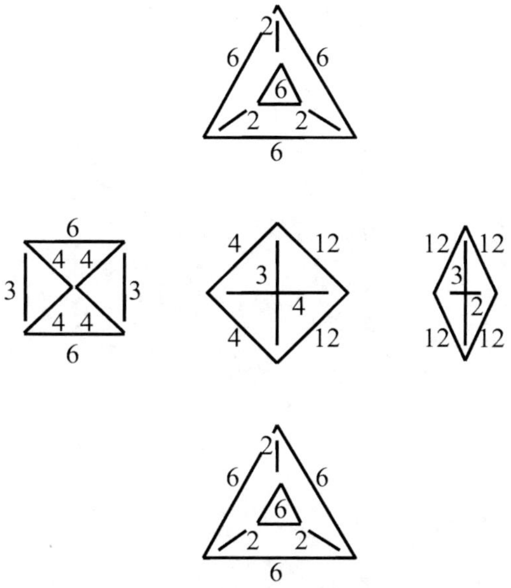

Fig.75: Incomplete [6-6] Family Grid

CIRCLE PACKING

We have seen that the degenerate skew polyhedra produce surfaces that are essentially restricted to two dimensions, albeit with a limited extension into the third dimension.

The spaces contained follow the lines of the three regular tilings of square {4,4}, triangle {3,6} and hexagon {6,3} and the polygonal clothing fits in and around those tilings when considered as the higher dimensional polytopes {4,4,2}, {3,6,2} and {6,3,2} respectively.

As with the 3-dimensional skew polyhedra, these internal lattices can be considered as joining the centre points of packed spheres, which restricted to the plane can be considered simply as packed circles.

Figures 76, 77 and 78 adjoining show the relevant packings and lattice grids that fit with the [4-4], [3-3] and [6-6] connected families.

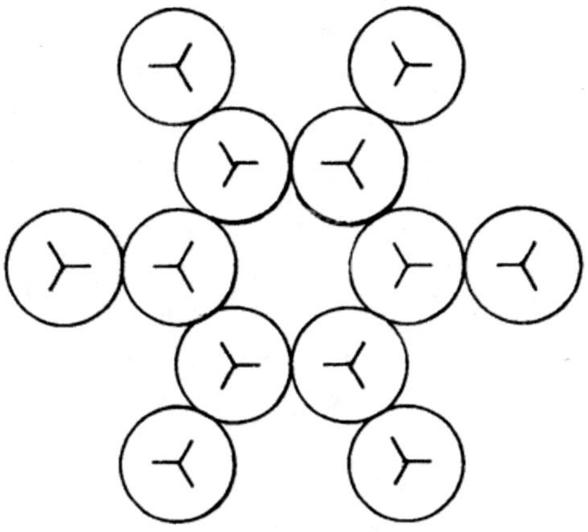

Fig.77: [3-3] Circle Packing

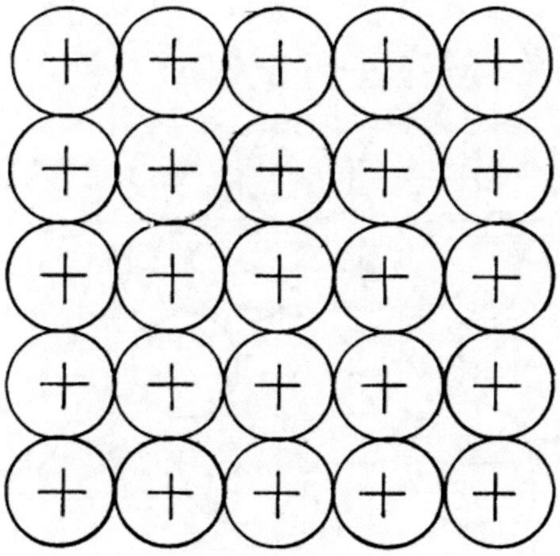

Fig.76: [4-4] Circle Packing

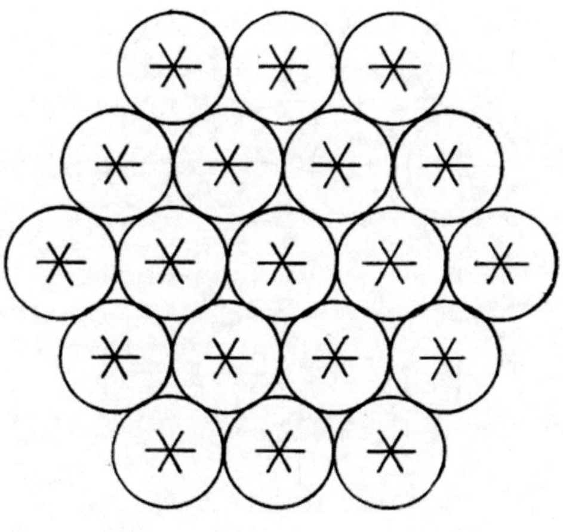

Fig.78: [6-6] Circle Packing

MORE CIRCLE PACKS

As with the 3-dimensional skew polyhedra, these internal grids can be expanded by the insertion of additional 2-connected nodes which result mostly in less dense packings.

Figures 79, 80 and 81 adjoining show the expanded packings and lattice grids that fit with the [2-4], [2-3] and [2-6] connected families, our original {4,8}, {4,6} and {4,12} degenerate skew polyhedra.

One further connecting lattice form needs to be illustrated, that of the [3-6] connected family. This is shown as a circle pack with connecting lattice in Figure 82 opposite.

The connections here form a {3,6} rhombic lattice, which has a parallel in the already seen {4,4} one that was the equivalent for the [4-4] connected family.

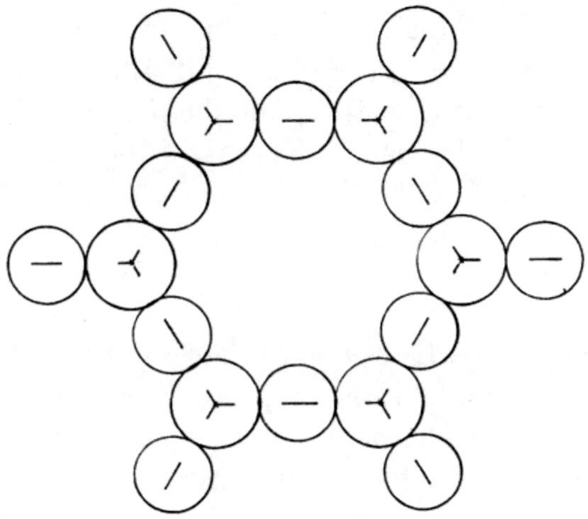

Fig.80: [2-3] Circle Packing

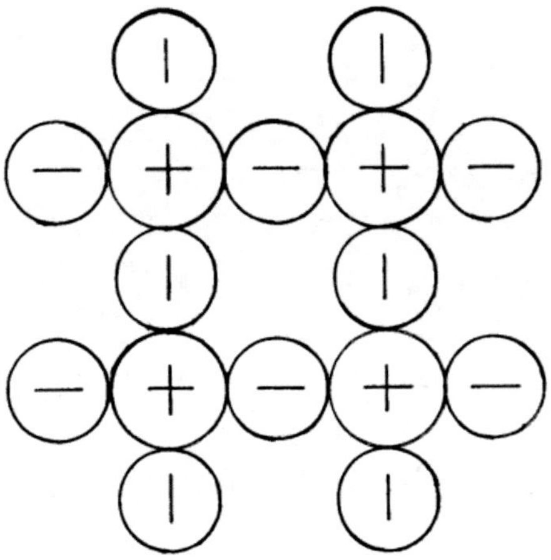

Fig.79: [2-4] Circle Packing

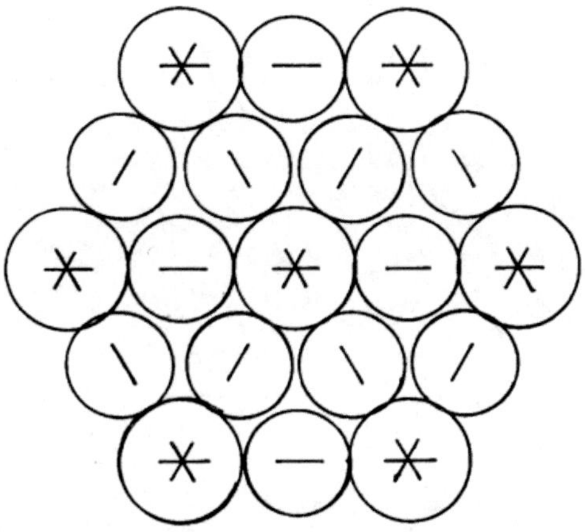

Fig.81: [2-6] Circle Packing

DEGENERATES

The degenerate skew polyhedra we have found are summarised below in Table 30, mirroring to some extent those of the three dimensional skew polyhedra in Table 15.

The full connectivities of these should include [2-∞] as the paired figures used thus far only include the spaces we consider inside each degenerate. The complementary space for these, that outside, has an infinity of 2-connections passing through the degenerate from a space either side.

The name 'degenerate' relates to both the fact that these do not venture fully into three dimensions and to the property of one group of them, which have vertices containing the same polygons as the hexaskews, but folded back to extreme positions.

Consideration of this folding back property leads to speculation whether there may be intermediate forms of these vertices. There are such, but they cannot be realised in three dimensions.

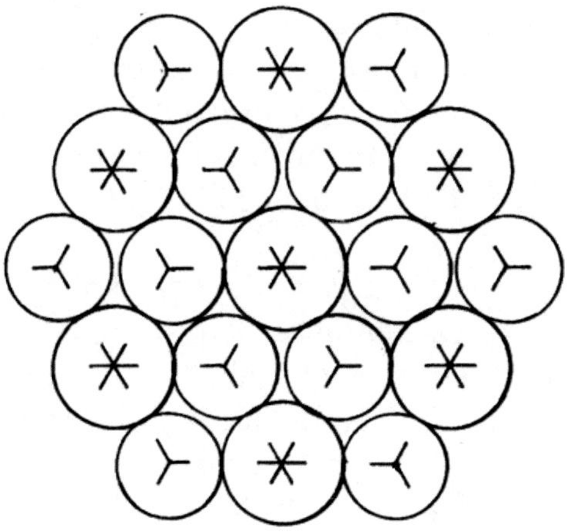

Fig.82: [3-6] Circle Packing

Centre vertex	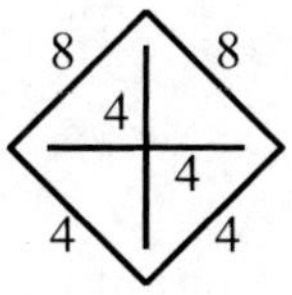	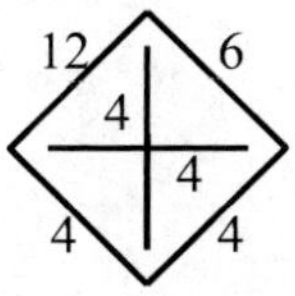	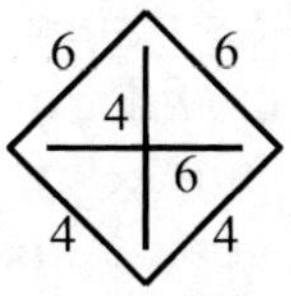	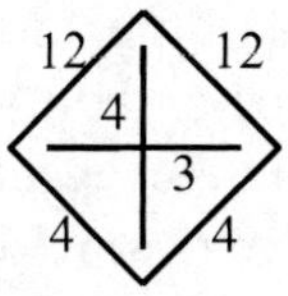
Truncated:	Tetratetraprism	Trihexaprism	Triprism	Hexaprism
Tetraskews	[2-4,2-∞]			
Tetraskews	[4-4,2-∞]			
Triskews		[2-3,2-∞]	[3-3,2-∞]	
Hexaskews		[2-6,2-∞]		[6-6,2-∞]
Trihexaskews		[3-6,2-∞]		

Table 30: Degenerate Skew Families

SUMMARY

Despite starting with but three regular skew polyhedra, those published by Coxeter in 1937, we have certainly expanded the field to present a far more complete picture of possible forms in three dimensions. These three originals have by the process of vertex transformation produced a range of intermediate semi-regular skew polyhedra, with more than one type of regular polygon at each vertex, all maintaining the same basic fourfold or sixfold connectivities.

Turning around the central tetrahedral vertices of these families we then found additional families and further tetrahedral vertices, all of which produced further semi-regular skew polyhedra by the process of vertex transformation. The only halt to the procedure was found to be when an odd valued polygon appeared. Such a polygon could not be obtained by truncation and led us into an incomplete family grid.

The spaces carved out by the various families of skew polyhedra have inter-connections that follow the lines of the lattices of the various crystal systems. They also weave their surfaces in and around the possible closepackings of polyhedra, effectively providing polygonal tilings of various minimal surfaces.

Overall we found five basic connectivities, each of which can be represented by a lattice model of sticks connected to nodes. We have identified the tetraskews with the minimal D-surface and the hexaskews with the P-surface and it seems likely that the octaskews, dodecaskews and icosaskews will have similar related minimal surfaces.

The degenerate skew polyhedra have also proved of interest, weaving their perhaps less complicated ways around the three basic tiling lattices.

Allowing for a fourth dimension, we will in the next and final volume of this series, be able to examine the hyperskews and their degenerates. These will wrap around the hypersolids and polyhedra in the same way that the skews and their degenerates have wrapped around the closepacks and tilings.

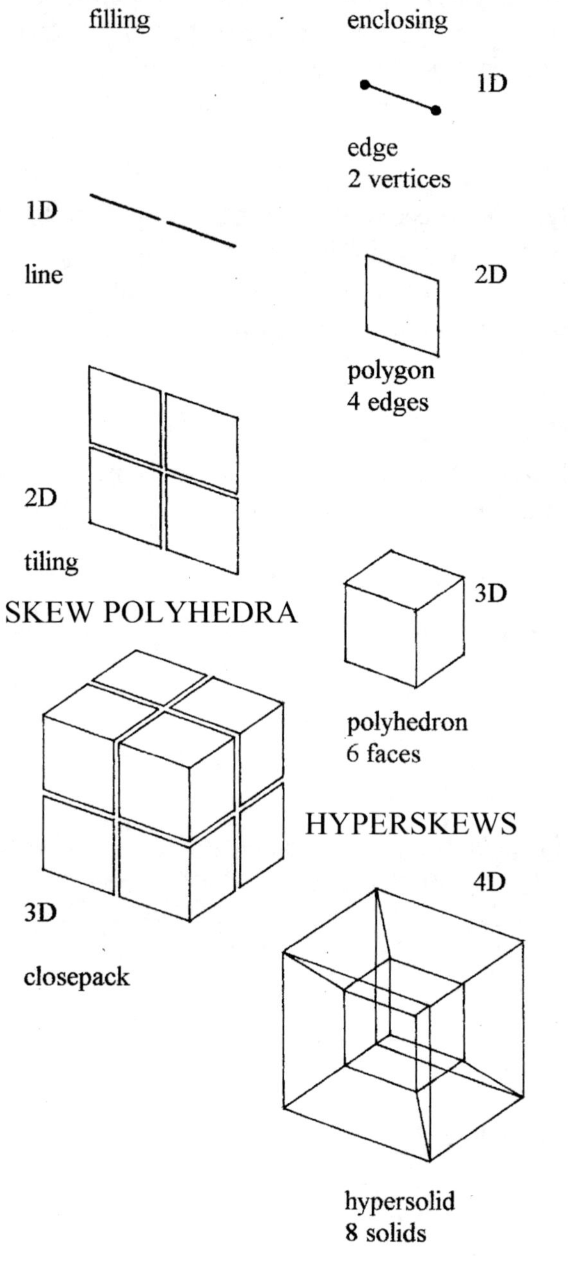

Fig.83: So What Follows?

BIBLIOGRAPHY

Abbott, E A 1962
Flatland
Blackwell

Aste, T & Weaire, D 2000
The Pursuit of Perfect Packing
Institute of Physics

Coxeter, H S M 1937
Regular Skew Polyhedra
Proc. London Mathematical Soc. p33

Coxeter, Longuet-Higgins & Miller 1953
Uniform Polyhedra
Phil. Trans. R. Soc. Lond. A246, p401

Coxeter, H S M 1963
Regular Polytopes
Macmillan

Coxeter, H S M 1991
Regular Complex Polytopes
Cambridge UP

Critchlow, K 1969
Order in Space
Thames & Hudson

Cromwell, P R 1997
Polyhedra
Cambridge UP

Grünbaum, B & Shephard, G C 1989
Tilings and Patterns: An Introduction
Freeman

Holden, A 1971
Shapes, Space and Symmetry
Columbia UP

Kappraff, J 1990
Connections
McGraw-Hill

Skilling, J 1974
The Complete Set of Uniform Polyhedra
Phil. Trans. R. Soc. Lond. A278, p111

Taylor, P 1990
The Geometry Trip
Nattygrafix

Taylor, P 1997
The Complete? Polygon
Nattygrafix

Taylor, P 1998
Incomplete Tilings
Nattygrafix

Taylor, P 1999
The Simpler? Polyhedra
Nattygrafix

Taylor, P 2000
The Star & Cross Polyhedra
Nattygrafix

Taylor, P 2004
Closepacks & Quasi-Closepacks
Nattygrafix

Taylor, P 2006
Hypersolids & Quasi-Hypersolids
Nattygrafix

Wenninger, M J 1971
Polyhedron Models
Cambridge UP

Weyl, H 1952
Symmetry
Princeton UP

Williams, R 1972
Natural Structure
Constable

OF RELATED INTEREST:

INCOMPLETE TILINGS
by Patrick Taylor 1998

being the second part of several comprising the complete? polyhedra

presents the tilings as assemblies of polygons that fill the plane; includes truncations, snub tilings, quasitruncations; investigating in total four uniform and seven compound symmetry families using regular, star, cross, compound and double polygons

76pp A4 paperback including 80 line drawings

ISBN 0 9516701 3 1 £6.00

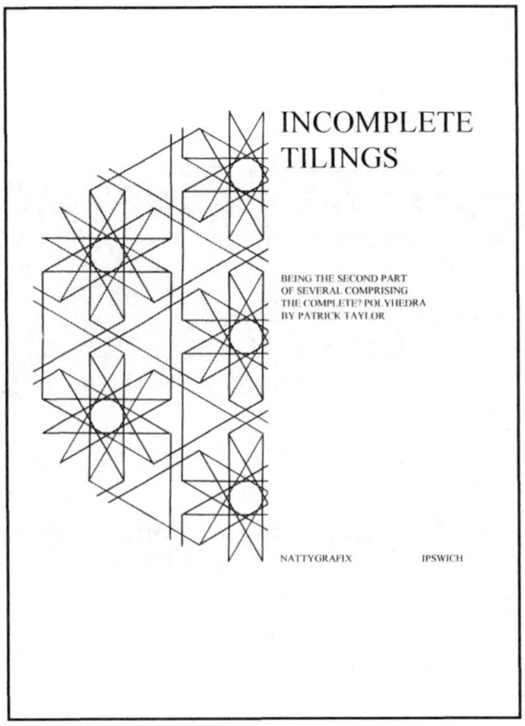

CLOSEPACKS & QUASI-CLOSEPACKS
by Patrick Taylor 2004

being the fifth part of several comprising the complete? polyhedra

presents the closepacks as assemblies of polyhedra that fill three dimensions; includes truncations and quasitruncations; investigating fully the {4,3,4} symmetry family, using the regular polyhedra of the {3,3} and {3,4}-{4,3} symmetry families

72pp A4 paperback including over 86 line drawings

ISBN 0 9516701 6 6 £6.00

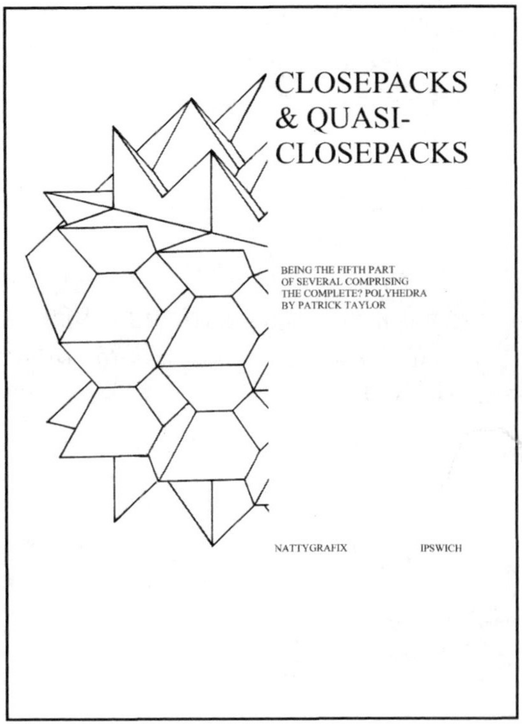